Geometric
Constraint Solving
and Applications

Springer-Verlag Berlin Heidelberg GmbH

Beat Brüderlin
Dieter Roller

Editors

Geometric Constraint Solving and Applications

With 127 Figures

Springer

Beat Brüderlin

Technische Universität Ilmenau
Lehrstuhl Graphische Datenverarbeitung
Helmholtzring 9, 98693 Ilmenau
Germany

Dieter Roller

Universität Stuttgart, Institut für Informatik
Breitwiesenstraße 20-22, 70565 Stuttgart
Germany

CIP-data applied for

Die Deutsche Bibliothek - CIP-Einheitsaufnahme
Geometric constraint solving and applications/ed.: Beat Brüderlin; Dieter
Roller. - Berlin; Heidelberg; New York; Barcelona; Budapest; Hongkong;
London; Mailand; Paris; Santa Clara; Singapur; Tokio: Springer, 1998
ISBN 978-3-642-63781-0   ISBN 978-3-642-58898-3 (eBook)
DOI 10.1007/978-3-642-58898-3

Originally published by Springer-Verlag Berlin Heidelberg New York in 1998

Cover Design: Design & Production, Heidelberg
Typesetting: Camera ready by authors

Printed on acid-free paper   SPIN 10675190   33/3142   5 4 3 2 1 0

Preface

This book is based on lectures presented at the first international Workshop on Geometric Constraint Solving and Applications, held at the Technical University of Ilmenau.

Experts from academia and industry were invited and reported on selected and relevant topics in this area. This volume is divided into four chapters, focusing on the semantics of constraints and data exchange, the role of constraints in conceptual design and collaborative design, new constraint solving mechanisms and representations, and constraints for freeform surface definition.

The first chapter presents a survey on the role of constraints in design. The section by Klein describes a high-level interface between constraint solvers and design expert systems of the future. In his section, Pratt reports on the progress and arising problems in the current effort of defining STEP, a data format for exchanging design data between remote sites and heterogeneous systems.

The second chapter includes new advances in constraint-based interfaces to conceptual design and collaborative environments. In the first section, Roller and Eck propose an active semantic network approach as a mechanism for constraint solving. The section by Anderl and Mendgen describes a system that helps designers cope with the complex interdependencies between features of an already advanced design, in a more intuitive way. The section by Csabai, Taiber and Xirouchakis introduces the idea of design spaces as an abstract interface between designers in a collaborative environment. In their section, Noort, Dohmen and Bronsvoort describe methods for detecting over- and underconstrained situations and suggest ways of how to notify interactive users in such cases. In the last section, Döring, Michalik and Brüderlin describe a shape modeling system, combining geometric constraints with topological constraints and Boolean set operations.

The third chapter proposes new representations for constraint-based design, as well as new approaches for solving constraint problems in specific domains. The first section by Rosendahl and Behrling describes the concept of generalized segments as an extension to dependency graphs by typical programming language constructs, and a degree-of-freedom-based method to arrive at such a directed graph from a given constraint problem. In their section, Hoffmann, Lomonsov and Sitharam extend previous graph-based approaches by combining flow-based degree-of-freedom analysis and generalized clustering. The section by Lhomme, Macé and Kuzo describes a new constraint solving mechanism in the projective geometry domain, based on Caley algebra, and its application to sketch interpretation. The next section, by Mathis, Schreck and Dufour describes a blackboard mechanism as an approach for integrating heterogeneous constraint solvers. In the section by Lamure and Michelucci, the combination of a graph-based degree-of-freedom analysis and iterative methods is described. The section by Solano and Brunet describes the use of relaxation to solve geometric constraint problems.

The fourth chapter includes new ideas for extending geometric constraint solving to the domain of freeform surfaces. Hagen, Heinz, Thesing and Schreiber describe the application of constrained optimization for achieving continuous boundaries and other properties between patches, in the context of reverse engineering of freeform surfaces from point clouds. The section by Loos and Greiner describes a constraint-based approach to freeform surface synthesis with optimal smoothness and symmetry properties.

Many people have contributed to the success of the workshop and to the preparation of this book. Particularly, we would like to thank the authors who submitted papers and prepared them for this edition. Special thanks also go to the reviewers and Michèle Johnson for their efforts in setting the standard for the quality of contributions in this volume.

Ilmenau / Stuttgart   Beat Brüderlin
Spring 1998   Dieter Roller

Contents

Chapter 4
Constraints for Freeform Surfaces   271

Chapter 1

The Semantics of Geometric Constraints

Overview

The Role of Constraints in Geometric Modelling

Rüdiger Klein[1]

Geometric constraints are important to increase expressiveness and flexibility of geometric and feature modelling. But other forms of consistency information are essential here, too, e.g., topological conditions, type constraints, etc. In many applications, overall consistency means to integrate also *logical conditions*. Due to the "built-in semantics" of geometry all these different forms of consistency information are not independent: they interact in various ways. Current geometrical modelling approaches do not provide the necessary expressiveness and flexibility. What is needed is an *expressive* and *integrated* way of geometric modelling. The G-Rep approach has been developed as a completely declarative, integrated approach to geometry modelling which allows a *tight conceptual coupling* with the various consistency representations. We describe the basic approach to declarative geometry modelling with constraints in G-Rep, and discuss some of the requirements and limitations.

Introduction

Though quite complicated, geometric modelling as applied in current CAD systems seems to be now one of the best understood fields of information processing in industrial environments [Requicha97]. Having (mainly) solved this part of the problem, the focus of attention shifts to new challenges — partially in response to the experiences gathered with current geo-

---

1. Daimler-Benz AG, Knowledge Based Systems Research Group, Berlin

metric modelling techniques, partially in order to fulfill new requirements. These new challenges can be summarized as follows:

- *More expressiveness*: Whereas in current CAD systems many geometry modelling operations have to be done step by step (which often is inefficient, tedious, and error-prone), more complex modelling facilities are needed. This is one of the motivations behind feature technology. But the modelling power of current feature approaches is too restrictive. Especially, intensional representations should be possible: a feature or 'piece of geometry' has to be represented in a *generic* way (independent of any concrete instantiation). In order to capture the intent of this generic 'piece of geometry' we need an adequate representational power. The various *geometric entities*, their *relations* and *constraints* must be represented explicitly.

- *Conceptual and system design*: Whereas current CAD systems are mainly dedicated to *component* shape design, in many applications the *interrelations* between *systems* and their *components* are a main issue. Conceptual design, preliminary design on higher abstraction levels, concurrent engineering taking down-stream aspects into account, etc. need support by next-generation CAD systems.

- *Interactivity and flexibility*: Design and engineering will continue to be processes in which humans play the key role – supported by more and more sophisticated computer systems. Due to the complexity of many design and engineering tasks problem solving must be incremental and iterative (including revisions and variants and their evaluation), and distributed between cooperating partners. Besides powerful human-computer interfaces this mainly implies having CAD and geometric modelling systems which allow *incomplete* specifications, *revisions* of previous decisions, the formulation of *relations and constraints*, and the *propagation* of changes and revisions.

- *Knowledge integration*: One way to get more powerful support from CAD systems to human users is the integration of various forms of knowledge into the CAD system.

- *Retrieval, reuse, and adaptation*: Today many design tasks start from scratch – simply because retrieval of previous cases from drawing repositories or from more or less unstructured CAD data libraries is too complicated. In the future the *retrieval*, *reuse*, and *adaptation* of previously solved design tasks will be essential. Cooperative design may include varying partners who have to exchange specifications, fit their designs to user requirements, reuse and adapt older designs to meet new requirements, etc. Therefore much more information capturing design intent, decision rationales, various forms of constraints, conflict resolutions, etc. have to be handled in conjunction with "pure geometry".

As a result, a completely different approach to geometric modelling will be needed: more expressive, more flexible, and better integrated with "other" aspects of information technology. Many of the "ingredients" needed therefore exist already (at least to some extent). One of the problems is to bring all these quite different parts together. Especially the geometric modelling kernel itself has to be modified in order to enable explicit referencing of the various geometric entities and their relations, to support revisions and changes, etc. Partially, this is a software engineering problem: some of the current geometric modelling kernels still include software written in Fortran sometime in the seventies. But it's a conceptual problem, too: neither CSG's history-based geometric modelling nor boundary representations allow us to fulfill all the representational needs. To capture design intent, to describe generic geometric shapes, relations and constraints between shapes, and to include various forms of knowledge about geometry a new representation of geometry itself is necessary.

Not all questions connected to these problems can be answered here. The following issues will be dealt with: first, from the quite abstract requirements catalog given above a more detailed and more concrete desription of geometric modelling needs will be deduced. Geometry representation is one aspect – the other, equally important one is geometry *generation*. In the following section we discuss our view on geometry generation. From this viewpoint, geometry generation can be seen as the incremental specification of new geometrical requirements the generated geometry has to fulfill. An expressive, integrated, sequence independent approach to geometry modelling is needed to meet these considerations. Essentially, it will result in the concept of an integrated and declarative approach to geometry modelling. The G-Rep concept [Klein96] has been developed to fulfill these requirements. It will be outlined in this chapter. Afterwards we describe the integration of various forms of logical reasoning and constraint techniques in G-Rep. Finally, we discuss some of the open research issues.

Expressive and declarative geometry modelling

CSG as well as boundary representations allow for a correct and complete representation of geometric shapes. From a *given concrete shape* in CSG or boundary representation a "sophisticated reasoner" (human or software) can retrieve, *in principle*, every shape-related information he/she/it is interested in. The "only" problem is that this may be very complicated – at least for a software system with *current* reasoning capabilities. The reason is that in each representation essential parts of the geometric information are left *implicit*: in CSG all surface, line, or topology information has to be calculated from the CSG tree; in B-Rep the volumetric information has to be derived form the boundary.

But not only retrieval of geometric information from a concrete geometric shape representation is needed – also the *generation* of new geometry (in order to satisfy new requirements), to guaranty the fulfillment of certain consistency conditions, various revision operations, and the propagation of changes resulting therefrom. Performing such operations on geometric representations in which essential parts of the information are left implicit seems intractable.

Requirements to an integrated geometric modelling

Consequently, a key requirement for an improved geometry representation is explicitness: all relevant geometric information should be explicit, and it should be represented in an integrated way. All geometric entities, their types, attributes, and relations should be represented:

- volumes (or cells),
- the related generic surfaces,
- the generic lines resulting from their interaction,
- the different topological entities which result from the interaction of all geometric elements,
- the relations of these topologoical entities to 'their' generic counterparts, etc.
- All these elements are characterized by their types (of different dimensionality: volumetric types such as cylinder, box, sphere; surface types

as planes, quadric or cubic; lines such as straight lines, parabolas, circle arcs; points).

- they have parameters and attributes (geometric or non-geometric in nature), and

- they are related or constrained in various ways.

Sometimes, besides the real geometric shapes also auxiliary geometric entities are needed (reference planes or lines, etc.).

Of course, the entities in a geometric model described in this way are *not independent*:

- surfaces are related to volumes,

- lines to the surfaces from whose intersection they result,

- topological entites are related to their "carrying" generic surfaces and lines,

- faces are related to their bounding edges, edges to their vertices, etc.

Representation of concrete geometric shapes

Taking general knowledge representation techniques we can fulfill these requirements for explicit and integrated geometric modelling of a concrete geometric shape in the following way:

- each geometric entity (independent of its kind and dimensionality) has to be represented by a *unique identifier*[2];

- all generic types (volumes, surface types, etc. including user defined ones) have to be represented by a concept, a sort, a type, or simply a unary predicate (depending on the concrete formalization chosen);

- the attributes and relations between geometric entities will be represented by roles or binary predicates. Some of them which reflect the *inherent semantics of 3D space* should be "built-in": for instance, a built-in relation 'bounding-surface' between volumetric entities and surfaces can be used to represent the fact, that the volume is bound by the related surfaces. Or the built-in relation 'topological-instance' expresses that

2.   The related problem of unique or persistent naming will be discussed in a forthcoming paper.

a certain topological entity has the respective generic entity as "carrier". User defined relations may be used to reflect application-specific aspects.

Thus, from this perspective, a *concrete* geometric model can be seen as a set of entities (described by their types and attributes) and relations between them. The representational expressiveness needed to describe these things is quite limited. Taking, for instance, first-order predicate logic (FOPL) as representational basis, such a geometric shape can be described by a set of ground (variable-free) atomic formulae of unary and binary predicates [Klein 96a]. Of course, the internal representation of a geometric model does not necessarily have to reflect these considerations. As in current geometric modelling systems much more efficient, special-purpose representations can be used. The main point is that the geometric modeller can *provide* this knowledge to a human expert or a knowledge based system in the way described.

Fig. 1 contains a simple example: a geometric entity 'b1' of (user-defined) type 'body', related to two plane-type entities 'p1' and 'p2' by built−in 'bound' relations. This entity b1 contains another object 'c1' of built-in type 'cylinder.' 'Contains' is also a user-defined relation with a defined semantics. Of course, there are many geometric entities and relations in this example (which, maybe, are not relevant for a human user or a knowledge based system).

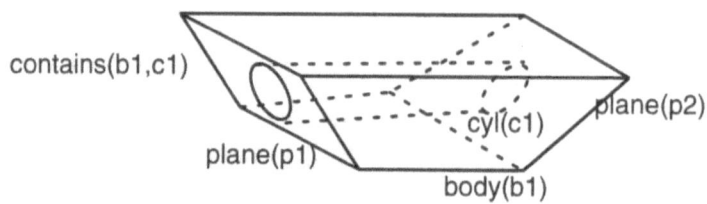

Fig. 1: A simple example of geometric entities

Due to the given semantics of the 3D space (and of geometric shapes) expressed by Euclidean geometry, this geometric representation (i.e., set of logical expressions) has to satisfy certain conditions in order to be consistent. Here consistent means to decribe a sound geometric shape in a unique, non-ambiguous way [Requicha 80]. For instance, type constraints exist between types of volumetric objects and surrounding boundary faces, or topological conditions have to be fulfilled. Again, these conditions do not necessarily have to be represented explicitly in a first-order language. Their fulfillment can be achieved by internal procedures of the geometric modeller.

Representation of generic geometric shapes

Two main problems result from these considerations:

- First, describing a concrete geometric shape (a solid or a feature) is one thing – also important, and more complicated, is the problem of *general* or *generic* descriptions of geometric shapes. Only this will give us the needed expressive power, flexibility, etc.

- Second, how can we *relate* such a generic shape description to a *concrete* solid or a concrete feature (feature instantiation or recognition).

The problems with generic shape and feature definitions come from the necessity to describe the set of geometric entities and features, which will be involved in the generic definitions, in an *intensional way*. This is also true for the conditions these entities and their relations have to fulfill. As a consequence, we need a *more powerful representation* then for concrete geometric shapes and features.

Categories of constraints

The following categories of constraints should be available in order to represent all relevant conditions for generic geometric shape and feature descriptions:

- geometric and arithmetic *constraints* as used in [Owen91; EREP94; Hsu et al. 97; etc.] in order to express parameter conditions: tangentiality or orthogonality constraints between geometric entities, incidence relations, the sum of the lengths of two features has to be less than the length of a third feature, etc.

- existence or non-existence requirements: for instance, for each cylinder entity the corresponding surface and topological entities have to exist, or for a cylindrical through-hole neither the top nor the bottom face should be in the geometrical model;

- type constraints of geometrical entities: for instance, a cylinder entity and its related surfaces are type constrained (a cylinder surface and two planes, arranged in an appropriate way);

- topological relations and constraints between entities: adjacency, incidence or inside relations, toplogical connectedness or unconnectedness; and

- logical combinations of a variety of these conditions such as implications or logical equivalencies.

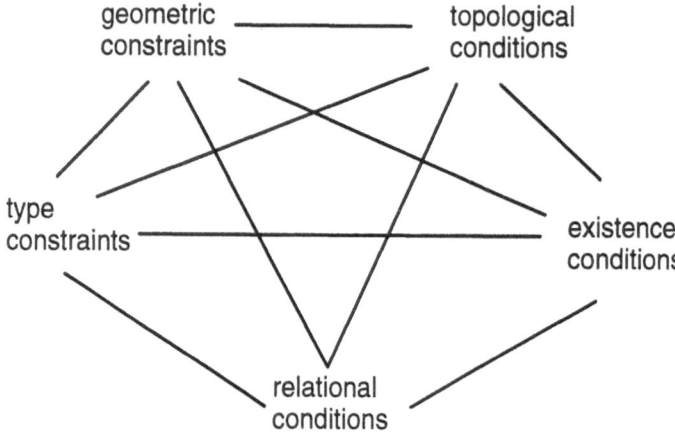

Fig. 2: the various constraint types relevant in geometric modelling and their mutual interactions

The explicit representation of all relevant geometric information (entities and their relations) and these five[3] kinds of consistency information result in an expressive geometric modelling approach. It would allow us to fulfill the requirements of an advanced geometric modelling described in the introduction.

The main problem is that constraints are not independent. They interact in different ways: a parameter change may cause a topology change, which results in the existence or non-existence of certain entities, modifies their types, establishes new relations, etc. (see Fig. 2).

---

3.   A special form of constraints combining type, existence and relational conditions are cardinality constraints [BL85], which may have some importance in geometrical modelling, too, and which need appropriate treatment.

How to use constraints in geometric modelling

Imagine a typical design scenario [Klein96]: at the beginning there is a (not necessarily complete) list of requirements the artifact to be designed should satisfy. Quite often these requirements are functional (i.e., they express the purpose the thing should be used for), but in some cases also structural aspects are specified. These requirements may be precise at different degrees, and they may possess varying priorities.

Functional requirements have to be transformed into structural descriptions which allow the artifact to provide the desired functionality. In the course of problem solving the functional requirements and the corresponding structural descriptions are worked out in more and more detail, until a level of abstraction is reached which is sufficient for the given problem. This refinement results in a description of the artifact, from which the fulfillment of the original requirements can be shown. At the same time, the general laws of nature, technical rules, normatives, etc. provide a set of consistency conditions the artifact has to fulfill, too. Due to the complexity of requirements and of the various conditions to be taken into account decision making in the design process may be quite complicated. It includes evaluations of recent states, revisions of former decisions, the creation of additional requirements (for instance, in order to fulfill a certain constraint), conflict resolutions including relaxations of previous requirements, and so on. This results in the typical incremental, *iterative* process of *hierarchical refinement*, *decision making* and *propagation*.

A formalization of the design process can be done in different ways. In [Klein96] we suggested the "General Design Engine" (GD_eE^4) as a logic-based approach to design. The key idea in this approach is to formalize the 'fulfillment' as *semantical entailment* ($\vdash$). Thus the requirements are described as a set G of formulae in an appropriate knowledge representation language (for instance, a modified first order predicate logic), and the general consistency conditions in the domain are represented as a set I of rules in the same language. Definitional knowledge D allows us to relate the more abstract notions in our domain to the more concrete ones in which the final solution will be described. Now the main point in the GD_eE approach is that the de-

---

4. not to be confused with the "General Diagnostic Engine" GDE of deKleer and Brown

signed artifact is represented as a semantical model M from which the original requirements and the consistency conditions are *semantically entailed*[5]:

$$M \vDash_D G$$
$$M \vDash_D I$$

Function to structure transformations can be complicated (even in relatively simple systems). So let's assume for the moment that these transformations are done by the human designer. Then we can focus our attention on structural descriptions, including geometric shapes. The definitional knowledge D contains, for instance, a set of generic feature definitions including existence conditions for those geometric entities which are needed to form the feature, relations between these entities, and geometrical constraints between their attributes in order to be consistent. The general consistency conditions I, for instance, can express that the wall thickness of any geometric entity should not fall short of a certain minimum value (in order to be mechanically stable or technologically feasible). Finally, the requirements G express that certain geometric features must exist which are related somehow and which have to fulfill certain geometric constraints.

Given these "ingredients", how can we come to a solution M of our problem? Two main issues are to be dealt with: constructivity and consistency. Constructivity means that for each single requirement $g \in G$ we need a procedure which generates a (partial) model M' from which g (using the definitional knowledge in D) is entailed. Depending on the kind of the requirement there can be one, a finite or an infinite number of such partial solutions. So, more precisely, this procedure should be able to generate *each* solution (maybe one by one). Overall consistency, of course, means that normally the various partial solutions generated to fulfill the different requirements are not independent. They influence each other in various ways, partially by explicit constraints defined between them, in part mediated by the semantics of 3D space. So the problem arises how to select those partial solutions for each of the requirements which are consistent all together.

In the currently used practice the human designer takes all requirements, tries to get an overview about existing principle solutions for each of them, analyses interdependencies between them, and then generates a solution step by step using the CAD system. Therefore he/she has to specify comple-

5.   for a more comprehensive formal discussion see [Klein96a] and also [KBN94].

tely the geometric modelling operations to be done. And even more importantly, the concrete way of geometric modelling applied in the used CAD system has to be taken into account very carefully. To a large extent *this* determines the sequence of modelling operations – often more than the inherent design problems themselves. Conflict resolutions and revisions may be very time consuming and need a deep understanding of the geometric modelling approach of the used system, too.

CAD systems with more expressive and flexible geometric modelling (as described in the introduction) would greatly help to overcome these shortcomings. A key issue here is the integration of the various forms of constraints described in this section.

Declarative integrated geometric modelling

A more expressive and flexible geometric modelling is currently prevented by mainly two reasons: the procedural approach applied in current geometric modellers, and the loss of information accompanied with CSG or B-Rep. Thus, in order to get a more powerful approach to geometry modelling two main issues are needed:

* a declarative representation of geometry; and

* an integrated approach to geometric modelling including volumetric, surfaces, lines, topological entities and their respective relations.

In order to get the meaning (the "real" geometry) of the geometric representation in current, "procedural" geometric modellers the *sequence* of all modelling operations is essential. The G-Rep concept [Klein96] has been introduced as a *declarative* approach to geometric modelling where the represented geometry is *independent from the sequence* of modelling steps. This is an important pre-condition for a more expressive and flexible geometry representation. Especially, it enables us to represent *all relevant aspects* of a geometric modell in an integrated way: the volumes, the associated generic surfaces and intersection lines, and the resulting topology (including the reference of each topological entity to its generic "carrier").

In order to get these two important properties of declarative and integrated geometry modelling, G-Rep is based on a two-level description of geometry [Klein96]:

- basic volumes and

- G-Rep solids Γ.

A basic volume is the 'atomic unit' of geometry modelling in G-Rep. It may be a standard elementary geometry like box, cylinder, sphere, it may be composed in a pre-defined manner, or it may be "configured" dynamically during problem solving (using the logic oriented representation). Associated with each basic volume is a set of parameters (which can be variables, maybe related by constraints) and a set of surfaces which form – as in a boundary representation – a topology of faces, edges, and vertices.

More complex geometries can be described in G-Rep by *composition* of basic volumes. For this purpose the notion of a G-Rep solid Γ in conjunction with G-Rep's composition operation '$\otimes$' are introduced. The essential point in the G-Rep concept is that the *meaning* of a G-Rep solid (the geometry it describes) depends neither on the sequence of the basic volumes it "contains" nor on the "history" by which the solid has been built up from the basic volumes through successive application of the $\otimes$-composition operator.

For being *declarative*, this G-Rep $\otimes$-composition has to be commutative and associative:

$$\Gamma_1 \otimes \Gamma_2 = \Gamma_2 \otimes \Gamma_1 \quad \text{and}$$
$$(\Gamma_1 \otimes \Gamma_2) \otimes \Gamma_3 = \Gamma_1 \otimes (\Gamma_2 \otimes \Gamma_3).$$

The way this can be reached is quite simple: instead of 'sequence' we take another kind of information which 'tells' the $\otimes$-composition operator how the individual basic volumes have to be combined.

By reasons of vividness we call this information 'density'. It can take any (positive or negative[6]) real number as value. A basic volume with a stronger (absolute value of the) density rules out those parts of its neighbour basic volumes with lower density values. A G-Rep solid thus is not a homogeneous thing but a conglomerate of different density zones.

---

6.   where a positive density means "material" and a negative density means a 'hole', thus enabling G-Rep a modelling power comparable to CSG union and complement

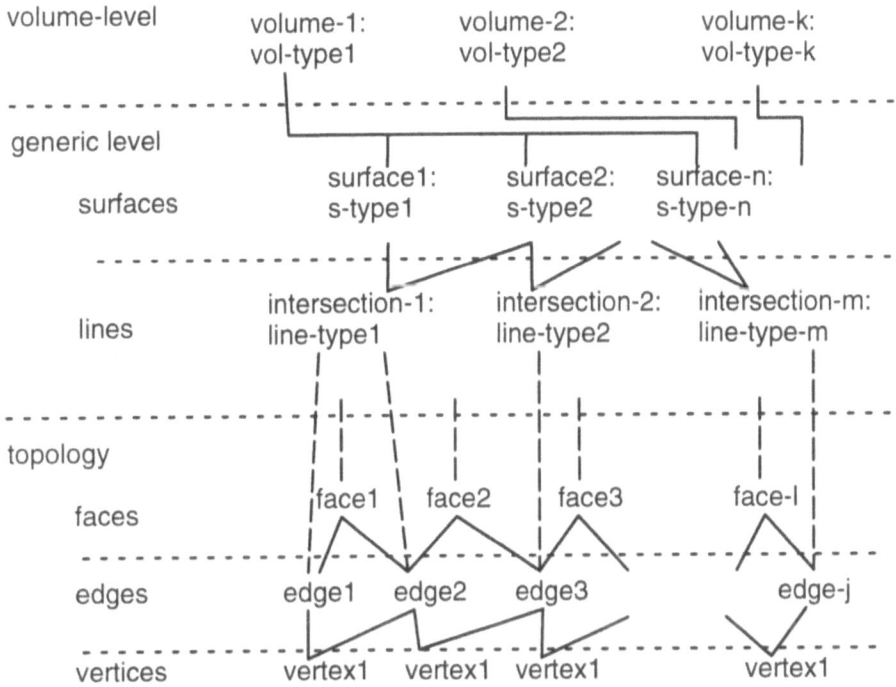

Fig. 3: Schematic view of the integrated tree-level geometry representatior

Volumes are associated with a (complete) set of boundary surfaces, each related to it by a 'built-in' bound relation. Each surface "knows" its intersection lines with the other surfaces, all surfaces and lines "know" their topological instances (indicated by dashed lines) which form a complete and unique boundary representation. Each element may be directly connected to any other by definition of relations.

This declarative way of geometry modelling provided by G-Rep allows us to integrate the different aspects of geometry representation: the various geometric entities and their relations. In G-Rep a three-level representation has been chosen for this purposes (fig.3):

- the top level description in G-Rep is defined by the basic volumes;

- each of these basic volumes is associated with a set of generic surfaces. This association may be represented by the built-in relation 'bounding-surface' or by more special (maybe user-defined) relations (like 'top' for the top-surface of a basic volume). The generic lines resulting from the intersection of the surfaces are part of this 'generic description level', too, and associated to their generating surfaces.

• The topology resulting from the density-controlled intersection of the various basic volumes and their generic surfaces and lines constitutes the third G-Rep representation level. All topological entities are described
  − in their topological relations to the other entities; and
  − as topological instances of their "generating" generic entities (or carriers).

The use of constraints in G-Rep

Generally, constraints can be used in two ways: they can be checked w.r.t. a given solution, or they can be used constructively to compute a solution. The first case is (with some restrictions) state of the art today: given the description of a geometric shape (in whatever form) and a set of geometrical or topological constraints [BB96; BDB96], it can be decided if these constraints are fulfilled by a given geometric shape or not.

For some kinds of constraints (for instance, geometric and linear arithmetic constraints) also the constructive approach is feasible [Owen91; EREP94; Hsu et al.97]: given such a set of constraints a solution or a set of them can be calculated which fulfill(s) these constraints, or it can be decided that no such solution exists for that set of constraints. In many applications it would be very desirable to have more powerful *constructive* approaches which allow the specification of type, relational, topological, etc. constraints from which a solution can be calculated. Leaving aside the principle limitations (as semi-decidability of first order logic) the question arises how current geometric modelling and constraint techniques can be improved to come closer to this general objective.

In the previous section we outlined our general view on the role of constraints in the design process. Constraints represent general "rules" of the domain as well as case specific needs. A model has to be generated as solution to a given design problem which fulfills these constraints. How this model generation can be done is the main point. Though this problem is currently not solved, we can show how the declarative and integrated way of geometry modelling provided by G-Rep can help in this respect.

$$\Gamma_1 = \{V_1, V_2\} \qquad\qquad \Gamma_2 = \{V_1, V_2, V_3\}$$

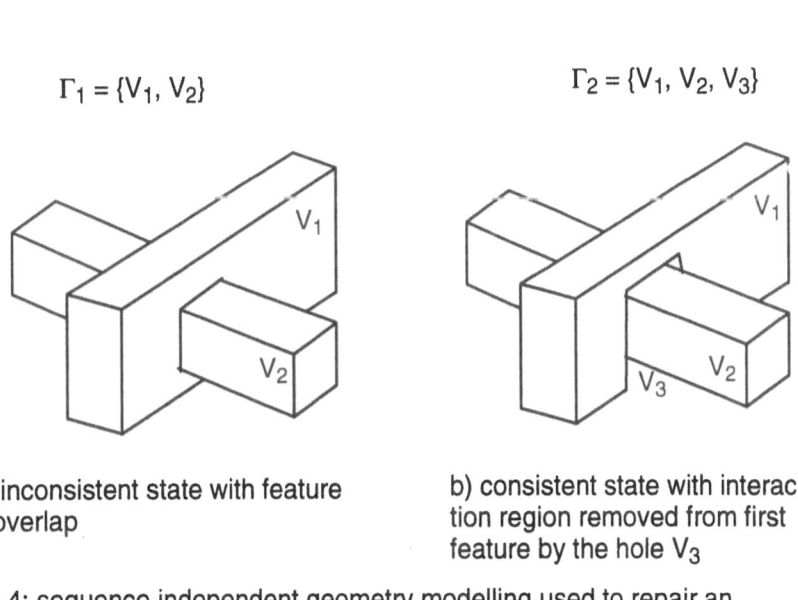

a) inconsistent state with feature overlap

b) consistent state with interaction region removed from first feature by the hole V_3

Fig.4: sequence independent geometry modelling used to repair an inconsistent state

The declarative, sequence independent way of geometry modelling in G-Rep enables us to perform the various geometry modelling steps in a way which follows the "inner needs" of constraint satisfaction and is not dictated by CSG or B-Rep modelling requirements. If, for instance, two features interact this means that in one of them a hole has to be cut to avoid this interaction. This can be done in G-Rep independent of the sequence in which these two features had been generated. The hole is simply generated, added to the G-Rep solid, and connected to the respective feature (see Fig. 4).

G-Rep's *integrated* approach to geometry modelling enables us to express the various kinds of constraints to the diverse entities (volumes, surfaces, topological entities, etc.) in a straightforward way. All entities and their relations can be adressed *directly* (see also Fig. 5). This is an important prerequisite for the usage of constraints in geometric modelling. It gives us the expressive means to represent the various kinds of constraints in a generic (as well as concrete) way:

- Each geometric entity has a type and the associated geometric parameters which are needed to describe this entity. This does not mean that in every G-Rep solid the parameters have to have exact values. The parameter values may be left variable, maybe constrained by geometric or arithmetic expressions. Type constraints can be given between entities.

- Because every geometric entity in G-Rep is represented explicitly, existence or non-existence constraints can be applied to a G-Rep solid [Klein96b].

- In the same way topological constraints can be used to express that topological relations have to exist or must not exist between certain topological entities.

- Geometric relations as adjacency or connectedness can be formulated between geometrical entities (as well as user-defined relations where the geometrical meaning of these relations can be formulated as logical constraints).

- Complex geometric representations can be used on top of a G-Rep solid in order to describe features. Due to the declarative and integrated style of G-Rep's geometry modelling arbitrary attributes, relations, and constraints can be used to define features [Klein97a].

Discussion and outlook

The G-Rep concept of geometric modelling has two main advantages: it is *declarative*, and it is *integrated*. On this basis a powerful and expressive, knowledge-based feature and geometry modelling concept with various types of constraints can be realized. In G-Rep features and geometry can be defined as generic and concrete entities using first-order predicate logic.

All geometric representational facilities provided by G-Rep can be integrated into these definitions. Though the declarative, integrated geometry representation of G-Rep provides us with the essential pre-condition for this expressive feature concept, this alone does not solve all problems.

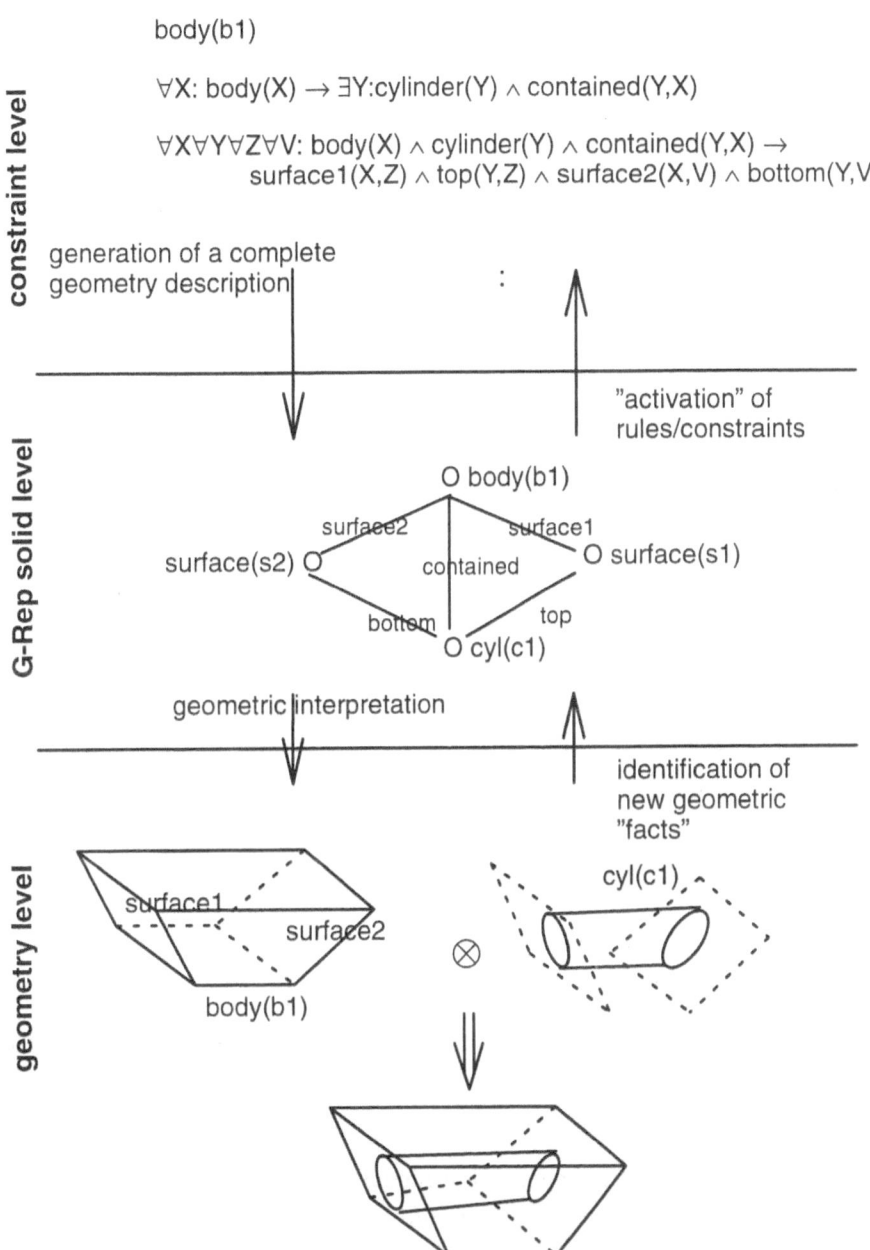

Fig. 5: A three level architecture of constraint processing in G-Rep.

Two main, closely related problems have to be solved in order to make practical use of G-Rep's principle advantages: solution generation and conflict resolution.

Solution generation: given a G-Rep solid Γ which was already generated to fulfill a certain set C of constraints and a new constraint c – how can we find all possible modifications Γ' of Γ which now fulfill $C \cup \{c\}$, too? – Conflict resolution includes conflict analysis (i.e., identification of those subsets of the constraint set which are not satisfiable) and appropriate "repairs", i.e., replacement of a partial solution by an alternative one.

The approach we are currently following is based on a three level system architecture, indicated in Fig. 5. On the top level the various types of constraints are managed: existence requirements, type constraints, geometrical relations, arithmetical and geometrical constraints, logical dependencies, and topological consistency conditions.

Constraint processing on the top level results in the generation of a G-Rep solid which fulfills all constraints. By the application of heuristic rules, defaults etc. it must be guaranteed that this solid is completely specified. Unspecified geometric parameters in more complex solids, for instance, could easily result in complicated reasoning processes to guarantee satisfaction of topological constraints.

From this second level the completely specified G-Rep solid is mapped to the "real" geometry on the third level.

Feature recognition done on this level may result in new constraints – propagated to the top level.

This system architecture is currently under implementation. It provides interesting means for expressive feature approaches as well as for a close coupling between geometry modelling and knowledge based systems.

Taking the complexity of geometric modelling, and taking the complexity associated with knowledge representation techniques (taxonomic reasoning, constraint processing, dependency maintenance), the question arises how all this can be made practical. Conceptual research as well as experimentation will be needed to get things ahead. One of the conclusions which can be drawn is that *all* these things will be needed "somehow" – having no taxonomic reasoning, no forward chaining, no arithmetic and geometric constraint solving, or no dependency maintenance would severely restrict the usefulness of the geometry modelling concept.

Consequently, research should be focussed on the following issues:

- expressiveness: what is really needed in order to represent the necessary feature concepts? For instance, what about negation?: under which circumstances which form of negation will be needed in feature representation?

- formalization: though the declarative and integrated geometry modelling of G-Rep is an important step towards a logic-oriented way of geometry representation, up to now there is no formal G-Rep calculus which allows us to make clear statements about the related knowledge representation and reasoning aspects.

- efficient algorithms and control: integration of the various reasoning techniques mentioned in a way which allows focusing on those reasoning activities which are really needed and avoids unnecessary efforts, i.e., a sophisticated control of the integrated reasoning techniques.

Acknowledgment

Lively discussion with researchers and practitioners greatly helped to develop the concepts described in this chapter. Special thanks to Frank Feldkamp, Michael Heinrich and Helen Leemhuis from our Knowledge Based Systems Research group.

References

[BB96]   Bidarra, R., Bronsvoort, W.: "Towards classification and automatic detection of feature interactions", 29th ISATA conference, Florence, Italy, 1996.

[BDB96]   Bidarra, R., Dohmen, M., Bronsvoort, W.: "Automatic detection of interactions in feature models", Proc. of the ASME Design Engineering Conference, Sacramento, California,1996.

[EREP 94]   Hoffmann, C.M., Juan, R.: EREP – an Editable, High-Level Representation for Geometric Design and Analysis, Techn. Report, Purdue Univ., West Lafayette, 1994.

[Gero95]   Gero, J. and Sudweeks, F. (eds.): Proc. of the workshop on
                formal design methods for computer-aided design, Tallinn,
                Estonia, 1995.

[Hoffmann 89]   Hoffmann, C.M.: Geometric and Solid Modelling: an Intro-
                duction, Morgan Kaufman, San Mateo, Calif., 1989.

[Hsu et al. 97] C. Hsu, G. Alt, Z. Huang, E. Beier, and B. Brüderlin: A Con-
                straint Based Manipulator Toolset for Editing 3D Objects,
                in: C. Hoffmann and W. Bronsvoort (eds.): Proc. of the 4th
                ACM Conference on Solid Modelling, Atlanta, May 1997.

[KBN 94]   Klein, R., Buchheit, M., and Nutt, W.: Configuration as Mo-
                del Construction: the Constructive Problem Solving Ap-
                proach, in: G. Gero and F. Sudweeks (eds.): Proc. Int'l. Conf
                AI in Design, Lausanne, 1994.

[Klein 96]   Klein, R.: G-Rep: Geometry and Feature Representation
                for an Integration with Knowledge Based Systems, Proc. of
                the IFIP 5.2 Workshop on Geomtric Modelling in CAD, Air-
                lie, Virginia, May 1996, Chapman Hall Publ., London,
                1996.

[Klein96a]   R. Klein: "A Knowledge representation perspective on geo-
                metric modelling", in: W. Strasser (ed.): Proceedings of the
                Blaubeuren−II conference on geometric modelling, Tübin-
                gen, Oct. 1996, Springer Verlag, Berlin, 1996 (to appear).

[Klein 96b]   Klein, R.: Towards a logical basis of design: the GD_eE ap-
                proach, in: F. Brazier, T. Smithers, J. Treur (eds.): Proc. of
                the workshop on logic-based approaches to AI in design,
                San Francisco, June 1996.

[Klein97]   R. Klein: "Dependency Maintenance in declarative geome-
                try modelling", 4th ACM Solid Modelling conference, At-
                lanta, Georgia, May1997.

[Klein97a]   R. Klein: "Towards knowledge based feature modelling,
                30th ISATA conference on Mechatronics, Firence,
                June1997.

[KRU94]   Krause, F.L., Rieger, E., Ulbrich, A.: Feature Processing as
                Kernel for Integrated CAE Systems, IFIP Int'l. Conf., Va-
                lenciennes, May 1994.

[Mäntylä 88]   Mäntylä, M.: An Introduction to Solid Modelling, Compu-
                ter Science Press, College Park, Maryland, 1988.

[Owen 91]   Owen, J.C.: Algebraic Solution for Geometryfrom Dimensional Constraints, Proc. ACM Solid Modelling Conference, Austin, Texas, 1991.

[Requicha80]   Requicha, A.: Representationd for rigid solids: theory, methods, and systems, ACM Computing Surveys 12:437−464, 1980.

[Requicha97]   Requicha, A.: private communication, 4th ACM Conference on Solid Modelling, Atlanta, May 1997.

[Rieger 95]   Rieger, E.: Semantikorientierte Features zur kontinuierlichen Unterstützung der Produktgestaltung, Hanser Verlag, München, 1995.

[Shah 93]   Shah, J.J.: Assessment of feature technology, J. for Computer Aided Design 23(93), 331−343, 1993.

[SHK93]   Salomons, O.W., van Houten, F.J.A.M., Kals, H.J.J.: Review of Research in Feature Based Design, Journal of Manufacturing Systems, 12/2, pp. 113 − 132, 1993.

[SMN94]   Shah, J.J.,Mäntylä, M., Nau, D.S.: Advances in Feature Based Manufacturing, Elsevier, Amsterdam, 1994.

Progress and Problems in Extending the STEP Standard

Michael J. Pratt[1]

The International Standard STEP (ISO 10303) is intended to facilitate the exchange of data between CAD systems. The first (1994) release of the standard allows the transfer of geometric product models in terms of geometry and topology plus configuration data (version numbers, etc). Since most CAD systems now allow the creation of variational models with parametrization, constraints and features, it is necessary to extend STEP to take into account these newer capabilities. This task is being addressed by the STEP Parametrics Group, working within ISO TC184/SC4. Progress on this task is reviewed as the first of the new standard documents begins to take shape.

Introduction

The development of the International Standard ISO 10303 (see Appendix), informally known as STEP (STandard for the Exchange of Product model data), started in 1984. The first version of the standard was issued ten years later, in 1994. The time taken reflects the fact that a major infrastructure was also created for what will be an ongoing development of STEP, possibly for many years to come. However, the ten-year gap and the fact that the nature of the standardization process requires the technical content of standards to be frozen well before their eventual release has led to a situation where STEP does not reflect the full capability of modern CAD systems. In particular, these systems can now generate parametrized, constraint-based and feature-based models [HD94], whereas STEP currently makes no provision for the representation of parametrized entities, constraints or features.

Several major components are therefore needed for the provision of the

---

[1]National Institute of Standards & Technology, Gaithersburg. MD 20899, USA. E-mail: pratt@nist.gov

missing parametric/variational capability in STEP. These will need to be embedded in a general framework, whose details are still being determined, although work is already in progress on some of the elements it will need to possess. The new developments are being pursued by the Parametrics Group within the standards committee ISO TC184/SC4 that administers STEP and several related standards falling under the general heading of 'Industrial Data'.

One major problem to be faced in extending STEP as described above is the dichotomy between explicit or declarative models and implicit or procedural models. The former are usually defined as boundary representations expressed in terms of faces, edges, vertices etc., and with all geometry explicitly specified in the model. The latter are the descendents of the constructive solid geometry approach to modelling, and in their case the model is defined in terms of the sequence of operations used in its construction. In this case there may be no explicit low-level geometrical or topological elements in the model. Most modern commercial CAD systems use some combination of these two approaches. For example, an extrusion feature is often defined in terms of an explicitly specified 2D closed profile and an implicit extrusion operation.

Release 1 of the STEP standard is designed primarily for the exchange of static (non-parametric) boundary representation models, which may include free-form geometry in the form of NURBS. With this capability, it can also handle, as special cases, simpler models such as wireframes, surface models without topology, faceted models and so on.

The technical content of any International Standard must be held static for some time well before its eventual release, and only editorial changes are allowed as the document passes from Draft to full International Standard status. Unfortunately for STEP, this phase in its development occurred at a time of rapid change in CAD systems, which is why it is now necessary to equip the standard with new capabilities. The explicit boundary representation models need to be supplemented by parametrization information expressing design freedom in the model, and with constraint information for the maintenance of design intent. Then, when a model has been transferred out of one CAD system into another it will be possible to edit it in accordance with the original designer's intentions. This will enable optimization of the design in the new system by appropriate adjustment of its parameters. Currently, only the basic Brep is transferred in a STEP exchange — all else is lost, and editing in the receiving system is consequently very difficult.

In fact, as pointed out by Rappoport [Rap97], pure Breps are nowadays used in most CAD systems as secondary representations for display purposes and to allow the user to pick model entities from the screen. The Brep has many other applications downstream of the design phase, but during the design process itself the primary representation used is generally based on the history of model construction. This permits editing in accordance with various aspects of design intent as built into the model during its creation. A pure Brep contains none of this information, as mentioned above.

For the transfer of Brep models, then, we know what information needs to be added to the STEP model representation. Once parametrization and constraint modeling mechanisms are in place we can use them to define features and transfer design feature information as well, which should be useful for many engineering applications subsequent to design.

However, the procedural or history-based aspect of the models generated by most modern CAD systems presents many more problems for STEP. For one thing, STEP currently has virtually no capability to capture this type of model except for a very basic CSG facility using non-parametrized volumetric primitives and the usual Boolean operations. The procedural design methods used by modern systems are based on the use of much more general and flexible operations, and so in this area it is necessary to start almost from scratch in defining a capability for the transfer of procedurally represented models. Further, because of the prevalence of hybrid models in practical systems this will have to work together with the extended Brep transfer capability.

Some of the difficulties associated with procedural modelling are well known, not least the persistent naming problem [CCH96, Kri95, Leq97, RS97]. However, in the STEP context there are also other problems of a related nature, and these will be discussed later in this section.

Constraint modelling in STEP

Some progress has been made in the modelling of constraints, but before this is described certain limitations on the scope of the work must be explained. These are dictated by the fact that the standard is intended to be independent of the native representations and algorithms of the sending and receiving CAD systems, whatever they may be. Thus the standard can only capture constraints in a *descriptive* manner — it will not transfer the actual variables used in solving those constraints, since they may differ between systems. The actual solution of constraint systems is also out of scope, for the same reason[2]. Fundamentally, then, what will

---

[2]This implies a need to distinguish the intended solution when multiple solutions arise. STEP will always transmit a fully evaluated model to be used for this purpose. The relationships between the elements in this model will be checked by the constraint solver in the receiving system; only minor adjustments should be necessary to ensure constraint satisfaction in the new computational environment, except in pathological cases.

be transferred is the set of geometric relationships existing in the model as perceived by the designer; these will include dimensional constraints and relationships such as parallelism and tangency. Both 2D and 3D constraints need to be captured, since some CAD systems now provide a capability for modelling 3D assemblies in which the parts are parametrically positioned and oriented and also appropriately constrained with respect to each other.

It is intended that the dimensional constraints available will include relationships involving algebraic or other relationships between parameters. This raises a problem in the STEP context because some of the basic resources in the standard were not designed for that kind of purpose. Specifically, STEP separates the information modelling level from the physical data transfer level, in the interests of possibly allowing different modes of data transmission other than (as at present) simple file transfer. Information modelling uses a language called EXPRESS, which is part of the standard. EXPRESS provides for the representation of mathematical expressions and functions, but for a rather specific purpose, namely the validity checking of entities which is carried out during translation from the sending system into the neutral STEP format ('preprocessing') and the subsequent translation into the receiving system ('postprocessing'). Thus, while expressions and functions can be formulated in EXPRESS they were not originally intended to be actually instantiated in the neutral file and transmitted into the receiving system, which is what is now required. In consequence, the fundamental descriptive methods of STEP are currently being extended to meet the new needs; these include the EXPRESS language itself (Part 11 of the standard) and the Physical File Format (Part 21). A further resource, the Standard Data Access Interface (SDAI) will shortly be added to the standard as Part 22, and this may also have to also have to be extended for the future.

At present all constraints except one (curve length — see the list below) apply to the pure geometric entities defined in Part 42 of the STEP standard [Int94a]; these are the usual implicit and parametric curve and surface types used in CAD modelling, including NURBS. Constraints are defined as entities for transmission in an exchange file, and each constraint entity represents an n-ary relationship between multiple geometric elements, e.g. a set of lines may be constrained parallel to a given reference line. Most of the constraint entities make provision for one or more optional reference elements; these must be specified in the case of a directed constraint, but will be absent if the constraint is undirected (e.g. when each line in a given set of lines is parallel to all the others but there is no precedence between them). The intended semantics, which take effect when the exchange file is reconstructed in a receiving system, are that in a directed constraint only the reference elements may be edited, but in an undirected constraint any of the elements may be edited. In either case, all the constrained elements should change to preserve the constraint.

The treatment of constrained dimensions is the subject of further work, since STEP already provides representations for dimensions, and the new capabilities under development must be compatible with the first release of the standard. However, there is a semantic difference between an ordinary dimension and a constrained dimension; the first may be edited, but the second should not be editable since its value, having been specified as fixed by the designer, presumably has some important functional significance. In the present version of the Explicit Geometric Constraints Schema dimensional constraints are defined as specialized subtypes of nondimensional constraints — the EXPRESS language is designed for the building of hierarchical relationships, and this was easy to do. However, it was found necessary to define two different types of distance constraints in order to be consistent with current engineering practice. All distances specified between lines on a drawing imply that those lines are parallel (in 3D the same implication holds for dimensions between planes). However, dimensions between points have no corresponding implication. Similarly, while the convention is that a dimension between a point and a line (or plane) is measured perpendicular to the latter, such a dimension implies no preexisting relationship between the elements concerned. Accordingly, the two types of dimensional constraint defined are (a) those involving parallel elements, and (b) those involving points.

The set of constraint types currently defined is as follows:

- parallelism — this has an undirected form and a directed form with one reference element. There is also a dimensional subtype, in which a constrained distance can be specified.

- point-distance — in the directed case the reference element may be either a point, line or plane. Multiple points may be constrained. In the undirected case the number of constrained points is limited to two, and a dimensional value is required.

- radius — this has a dimensionless form, 'the radii of all these arcs are the same', and a dimensional form, 'the radii of all the constrained arcs have the same specified value'.

- curve-length — asserts that the lengths of all members of a set of *trimmed* curves are equal. There is a dimensional form allowing the value of the length to be specified. At present this is the only constraint type operating on *bounded* geometric elements.

- angle — constrains a set of lines or planes to make the same angle with a reference element, or in the undirected case specifies the angle between precisely two such elements.

- direction — a vector-valued constraint used for constraining the directional attributes of linear elements such as lines or planes.

- perpendicularity — like all that follow, this type of constraint has no dimensional subtype. There may be either one or two reference elements (lines or planes), and all the constrained elements are required to be perpendicular to them. There is also an undirected form, in which two or three elements are required to be mutually perpendicular.

- incidence — this must have at least one reference element. In its simplest form, it simply asserts that one or more constrained entities are contained within that reference element. However, an inverse interpretation is allowed in which there may be multiple reference elements and they are required to be contained in the constrained elements. Thus the incidence constraint could be used to constrain a curve to interpolate a set of points.

- tangency — may be used to specify multiple tangencies between a set of reference elements and a set of constrained elements, for example two circles and two lines.

- coaxial — constrains a set of rotational elements to share the same axis or to share a specified reference axis.

- symmetry — constrains two ordered sets of elements to be pairwise symmetric with respect to a given line or plane.

- fixed — used to fix points and directions in absolute terms for anchoring local coordinate systems in global space.

Initially it was intended to incorporate sense comparisons in the parallelism and tangency constraints, to remove ambiguities. For example, there are four possible tangent configurations of a line with two circles (provided one circle does not contain the other), but if all the geometric elements have senses then a particular solution may be identified by specifying that the senses agree or are opposed at each of the points of tangency. However, the current inclination is to omit such comparisons. The reason is that the geometric elements defined by STEP may be used in various ways; for example, a trimmed surface may be created in terms of an underlying surface and a set of trim curves, and depending on the application the sense defined for the trimmed surface may be opposite to that of the underlying surface, so that senses used in disambiguating a tangency constraint on it may not have the desired effect. Sense comparisons have therefore been omitted from the new fundamental STEP resource described here; it is felt more appropriate that they should be defined at some level of STEP closer to that of the actual applications. Several such levels are available in the architecture of the standard.

The list given above may be modified in the light of feedback from CAD vendor organizations, and in particular is likely to be extended to handle constraints on free-form geometry.

The first five of the constraint types in the list given above have dimensional subtypes, as already mentioned. The values of dimensions in these cases may be specified either (a) numerically, (b) by reference to a dimensional attribute of a reference entity, or (c) in terms of an expression involving parameters or dimensional attributes, which is evaluated to give the desired value. The precise mechanism for doing this will depend upon the changes currently being made in the EXPRESS language for its Version 2 release, due in the year 2000 or thereabouts..

The ENGEN Project

The development of the Explicit Geometric Constraints Schema for STEP has been progressing in parallel with a Project Called ENGEN (Enabling Next GENeration design). This has had as one of its objectives a demonstration of the transfer of constrained models between different CAD systems. There has been significant liaison between the STEP Parametrics Group and the ENGEN Project, and both have benefited from the exchange of information.

One of the major deliverables of ENGEN has been the ENGEN Data Model, which makes provision for the capture of both explicit and unevaluated or history-based models. The corresponding STEP Parametrics document is known as the Parametrics Framework. Neither document is complete, and both of them have fleshed out the details more in the area of explicit geometric constraints than in any other area. Collaboration between ENGEN and the STEP Parametrics Group has led to considerable convergence between the constraint models in the two documents, though because of the limited scope of ENGEN there are a few notable differences:

- ENGEN is essentially restricted to 2D constraints, while STEP also allows 3D constraints

- STEP defines a smaller number of more general constraints

- ENGEN applies to a wider range of elements including trimmed curves

- ENGEN explicitly specifies 'sense' information

- ENGEN constraints have optional accuracy attributes

On the other hand there are strong similarities in other respects. Both models make use of optional attributes ('reference elements') to distinguish between directed and undirected constraints, both separate linear dimension constraints into forms involving only linear elements and forms involving points, and there is a very similar range of functionality.

The differences arise mainly because ENGEN had to produce working implementations in a limited time. This required a fairly narrow project scope, and also led to a rather different 'flavour' of data model. The ENGEN Data model is intended for direct implementation via translators to and from CAD systems. The STEP model, on the other hand, is less close to the implementation level; it is resource on which STEP Application Protocols (APs) will be based, and it is those APs rather than the resource model itself that will be the basis of practical model transfer. Thus the STEP model is defined at a more abstract level. Also, its geometric scope includes the whole of the STEP geometry resources, whereas the ENGEN model is based on a modified subset because corners had to be cut to get the work done in the time available.

The issue of accuracy of constraint satisfaction, mentioned above, needs further explanation. The kind of situation that will arise in practical model exchanges is that logical constraints will disagree with geometric data. For example, a constraint may specify that two lines are perpendicular when the explicit model transmitted together with the constraint set has those lines at an angle of 89.9999° because of rounding errors. This is in fact very closely related to a problem that already occurs in STEP data transfers; the geometry and topology of a model do not always agree. The topology may assert that two edges are connected when the geometry specifies two slightly different points for the corresponding endpoints of those edges. STEP currently assumes the transmission of ideal models, in which the geometry and topology of the model are totally self-consistent, but unfortunately CAD systems do not generate such models. At present this problem is being tackled mainly by the CAD system developers, who are striving to deal with their internal numerical tolerances in a more consistent way. In this author's view there is little point in building accuracy criteria into constraints in STEP when the closely related topological relations are not similarly furnished. A satisfactory solution needs to be found that will cover both cases.

The ENGEN demonstrations that were performed in the summer of 1997 involved three commercially available CAD systems, Computervision CADDS5, Parametric Technologies' ProEngineer and SDRC I-DEAS. The demonstration part was a connecting rod from a car engine, defined as an extrusion of a 2D profile containing several geometric constraints. It was shown that the part could be edited in the receiving system subject to the transmitted constraints. An account of this work is given by Shih and Anderson [SA97].

One interesting fact that emerged was that different systems take different views of the distinction between topology and constraints. Thus some systems regard all topological connections as individual constraints, while others deal with connectivity and constraints differently. This discrepancy gave rise to problems, as might be imagined, but ways were found of overcoming them.

The primary significance of the ENGEN work from the point of view of STEP Parametrics is that a proof-of-concept demonstration has been given of the practical exchange of geometric constraint data between systems, using a constraint information model very close to that being developed for STEP.

History-based Models

It is envisaged that the ability to transmit a boundary representation model with parametrized dimensions and explicit geometric constraints will enable the editing of the model in a receiving system provided that the changes are in some sense 'small'. One possible restriction might be avoidance of changes leading to topology modification in the model. But the necessary restrictions probably depend upon the precise nature of the receiving modeling system.

For full editability of models in a wide variety of different receiving systems the problem of transmitting procedural or history-based models will need to be tackled. This gives rise to several subsidiary problems, including the well-known persistent naming problem mentioned above. This may be illustrated for the case where the designer picks a model element from the screen display in specifying a reference entity for a constructional operation. If a procedural representation of the resulting model is edited and rerun, then the system has to identify the element in the modified model corresponding to the one originally picked by the designer. It is highly debatable whether this is always possible, but unless it can be achieved most of the time intuitive editing in a receiving system following a model exchange will frequently be impractical. For this reason, the STEP Parametrics Group will be devoting attention to mechanisms for persistent naming in the near future. Even if a satisfactory mechanism can be identified for use in STEP, there remains the problem of whether the different mechanisms used by commercial CAD systems will map satisfactorily onto it. On the other hand, the adoption of a satisfactory method in STEP may cause some CAD vendors to migrate towards the STEP solution if it is perceived to be better than their own current approach. This matter therefore needs very careful attention.

Apart from persistent naming there are several other problems associated with standardizing a representation for procedural modeling:

- The determination of an appropriate set of constructional functions [Pra98]. This is difficult because CAD systems differ quite widely in the functions they provide. One solution would be to translate all high-level operations into sequences of low-level operations on individual geometric or topological entities, but this would lead to very long exchange files and also make editing of the file for design optimization in the receiving system much more tedious. On the other hand, the transfer of only high-level operations is likely to lead to incompatibility because different systems implement different high-level operations or implement similar operations in different ways. There is thus a difficult problem of choosing the appropriate level of granularity for the standardized constructional operations. Hoffmann, in his EREP proposal [HJ92] appears to suggest that it is sufficient to provide a fairly limited set of constructional operations for parametrized design features, but there is considerable doubt as to whether this will be adequate.

- There are problems, related to the persistent naming problem, that arise in the context of instancing in assemblies and in referencing elements of components extracted from external libraries of standard parts. In either case it will be necessary to provide a means for making explicit references to elements of objects that are only described implicitly, as in the basic persistent naming requirement.

- Similarly, it is likely to be necessary to provide a means for referencing auxiliary constructional elements in a model that are used as reference elements in some operation but do not appear in the final model, perhaps because some subsequent operation has deleted them. A simple example is provided by a rectangle, subjected to a constrained dimension between two of its diagonally opposite corner points. If one of those corners is subsequently rounded the associated constrained point will disappear, though clearly its effect should still be present for design intent purposes.

- Finally, it will be necessary to integrate the history-based model representation capabilities with the explicit modeling capability of STEP, because many CAD systems use hybrid techniques. For example, 2D profiles are often constructed in terms of explicit lines and curves subject to geometric constraints, and then used in procedural operations such as extrusions or rotational sweeps.

One noteworthy point regarding constraints in history-based models is that they may be defined implicitly rather than explicitly. Consider, for example, a constructional operation for the generation of a line, parallel to an existing line at a specified distance from it. Here the parallelism is implicit in the constructional procedure. If a system subsequently generates an evaluated or explicit model from the history representation, the implicit constraint may or may not be transferred into the detailed model as an explicit constraint. However, the responsibility of STEP will terminate once it has delivered the procedural model; whether or not the explicit constraint is generated in the receiving system is the responsibility of that system.

Features and Assemblies

Work has not yet started on feature representations, because it is felt necessary to finish the definition of the parametrization and constraint representations first. STEP already provides certain limited feature capabilities, but primarily in a manufacturing context, and with no parametrization or constraints. The existence of these capabilities of course complicates the task of the Parametrics Group, since it has to ensure that any general feature modeling mechanism it develops for STEP is compatible with what has been defined previously. There are other areas of the work where this is also the case, notably (as mentioned earlier) in connection with dimensions and tolerances.

Assemblies is another topic that will be addressed in the medium-term future. Currently STEP's assembly modelling capability takes the form of a hierarchy based on assembly/subassembly/component relationships, which is adequate for generating parts lists and bills of materials. However, it is desired to add further detail to this model by including assembly features and mating relationships. The explicit geometric constraints schema will provide the necessary capability for constraining parts with respect to each other. The result will be an assembly model that can, for example, form the basis of kinematic simulations, something that is possible in many current CAD systems.

Summary

Since its initial release in 1994 the STEP standard has been the subject of widespread testing by industry. Early problems are rapidly being overcome. These were mainly due not to deficiencies in STEP but rather to deficiencies in the models generated by some CAD systems, often as a result of poor user practice, and of differences in internal numerical tolerances between CAD systems.

The standard is already in production use in some major companies. For example, Boeing uses STEP to exchange data with its three engine suppliers General Electric, Pratt & Whitney and Rolls-Royce. The purpose is to ensure that all four companies work with consistent geometry in the area of the engines and engine mounting struts of the 777 aircraft. Between them, the four companies use three different CAD systems. Current savings at Boeing due to the use of STEP for this limited application alone are reported to be well over $1M per year, and there are fewer delays in the communication of geometric data because the transfer is now largely automatic.

While STEP model exchanges of pure boundary representation models are meeting with increasing success, many companies feel the lack of the design intent data that allows easy modification of the model in the receiving system. This is particularly so in the automotive industry. The Parametrics Group within the STEP community is therefore working, as has been described, towards providing increased capability for Version 2 of the standard. The new facilities are intended to permit the capture and transmission of parametrized models with constraints and features, but there are significant problems to be overcome before this becomes a practical possibility.

Acknowledgements

The author would like to thank his collaborators in the STEP Parametrics Group, in particular Noel Christensen of Allied Signal, Akihiko Ohtaka of Nihon Unisys and Chia-Hui Shih of IBM, for providing insight into some of the topics covered above. Many others, too numerous to name individually. have also made valuable input to the work of the ISO Parametrics Group.

Disclaimer

Certain software companies and commercial software systems are mentioned in this contribution. Such identification does not imply recommendation or endorsement by NIST; nor does it imply that the products identified are necessarily the best available for their purpose.

References

[CCH96] V. Capoyleas, X. Chen, and C. M. Hoffmann. Generic naming in generative constraint-based design. *Computer Aided Design*, **28**, 1, 17 – 26, 1996.

[HD94] J. Hoschek and W. Dankwort, editors. *Parametric and Variational Design*. Teubner-Verlag, Stuttgart, 1994.

[HJ92] C. M. Hoffmann and R. Juan. EREP — An editable, high-level representation for geometric design and analysis. In P. R. Wilson, M. J. Wozny, and M. J. Pratt, editors, *Geometric Modeling for Product Realization*. North-Holland Publishing Co., 1992. (Proc. IFIP WG5.2 Workshop on Geometric Modeling, Rensselaerville, NY, Sept/Oct 1992).

[Int94a] International Organisation for Standardisation. *Industrial Automation Systems and Integration – Product Data Representation and Exchange, Part 42 – Integrated Generic Resources: Geometrical and Topological Representation*, 1994. (International Standard ISO 10303-42).

[Int94b] International Organisation for Standardisation. *Industrial Automation Systems and Integration – Product Data Representation and Exchange*, 1994. (International Standard ISO 10303, informally known as STEP).

[Kri95] J. Kripac. A mechanism for persistently naming topological entities in history-based parametric solid models. In C. M. Hoffmann & J. R. Rossignac, editor, *Proc. 3rd ACM Symposium on Solid Modeling and Applications, Salt Lake City, 17 – 19 May 1995*, pages 21 – 30. ACM Press, New York, NY, 1995.

[Leq97] R. Lequette. Considerations on topological naming. In M. J. Pratt, R. D. Sriram, and M. J. Wozny, editors, *Product Modeling for Computer Integrated Design and Manufacture*, pages

394 – 403. Chapman & Hall, London, 1997. (Proc. IFIP WG5.2 Workshop in Airlie, VA, 19 – 23 May 1996).

[Pra98] M. J. Pratt.   Parametric representation and exchange: Background study for procedural modeller interface.   ISO TC184/SC4/WG12 Working Document N106, 1998.

[Rap97] A. Rappoport. Breps as displayable-selectable models in interactive design of families of geometric objects. In W. Strasser, R. Klein, and R. Rau, editors, *Geometric Modeling: Theory and Practice*. Springer-Verlag, 1997. (Proceedings of the 'Blaubeuren II' Conference, Blaubeuren, Germany, October 1996).

[RS97] S. Raghothama and V. Shapiro.   Boundary representation variance in parametric solid modeling.   Technical Report SAL 1997 – 1, University of Wisconsin Madison, Spatial Automation Laboratory, 1997.

[SA97] C.-H. Shih and W. Anderson. A design/constraint model to capture design intent. In C. M. Hoffmann & W. Bronsvoort, editor, *Proc. 4th ACM Symposium on Solid Modeling and Applications, Atlanta, GA, 14 – 16 May 1997*, pages 255 – 264. ACM Press, New York, NY, 1997.

[USP96] USPro. *Computer Aided Processing of Engineering Drawings and Related Documentation (IGES)*. US Product Data Association, Fairfax, VA, USA, 1996. (IGES Version 5.3: American National Standard USPRO/IPO-100-1996).

Appendix: ISO TC184/SC4 and STEP

The International Organisation for Standardisation (ISO) administers a range of committees and subcommittees. ISO Technical Committee 184, Subcommittee 4 (TC184/SC4) is concerned with the development of international standards for the digital representation of product data and manufacturing management data. This is the forum in which the STEP standard (ISO 10303) is being developed. The first version of STEP [Int94b] was released in 1994. Earlier related standards (for example IGES [USP96]) were intended primarily for the exchange of pure geometric data between design systems, but STEP is intended to handle a much wider range of information covering the entire life-cycle of a product.

ISO TC184/SC4 is also responsible for the development of

ISO 13584 (Parts Library) – a future standard for making information in libraries of standard parts accessible to CAD system users,

ISO 15531 (MANDATE) – a future standard for non product-related manufacturing management data,

ISO15926 (Oil & Gas) – a future standard for data concerned with the life-cycle of oil and gas production facilities.

All four standards are based on a similar approach to data modeling, in particular the use of the EXPRESS information modeling language.

The STEP standard is being released in parts. The initial release contained twelve of these, and at the time of writing there are fifteen. Many more are in preparation, dealing with specific product ranges (e.g. automotive, AEC, shipbuilding, electrical,...) and different aspects of the product life-cycle (design, finite element analysis, process planning,...).

The structure of STEP is fairly complex. The lower Part numbers (100-series and below) define the infrastructure and a set of integrated resources. The actual data exchange standards are specified by Application Protocols in the 200-series, and these are defined in terms of the lower-level resources. Part 11 specifies the EXPRESS information modeling language, which is used for the formal definition of constructs in the exchange files. Part 21 defines the physical file format used in file exchanges. Part 22 defines the Standard Data Access Interface (SDAI), an access interface to STEP data in data repositories.

The released parts of STEP dealing with geometry transfer are three Application Protocols, AP201 (Explicit Draughting). AP202 (Associative Draughting) and AP203 (Configuration Controlled Design). The first is concerned purely with 2D drawing information. The third covers wireframe, surface and boundary representation solid models. The content of AP203 models is restricted to geometric and topological data, together with 'configuration' information relating to such matters as version control and release status. AP202 is a kind of hybrid permitting the transfer of CAD models with associated drawings.

STEP is designed to operate in the first instance as a 'neutral file' transfer mechanism. Each CAD system must be provided with a *preprocessor* and a *postprocessor*. Their functions are, respectively, to translate native data from the sending CAD system into the neutral STEP format, and to translate from the neutral format into the native format of the receiving system. This philosophy only requires the provision of $2n$ translators for exchange between any pair chosen from n systems, rather than $n(n-1)$ if 'direct' translators have to be written. As mentioned above, an alternative is to store STEP information in a database and share it through the use of the SDAI.

Many CAD vendors have developed or are developing STEP AP203 translators; some are already commercially available, while others are under test. Several third-party software vendors are also marketing STEP AP203 translators.

The currently released parts of the standard are

Part 1	Overview
Part 11	EXPRESS language (used in writing the standard)
Part 21	Physical file format
Part 31	Methodology and framework for conformance tools
Part 41	Fundamentals of product description and support
Part 42	Geometric and topological representations
Part 43	Representation specialisation
Part 44	Product structure configuration
Part 46	Visual presentation
Part 47	Shape variation tolerances
Part 101	Application resources: draughting
Part 105	Application resources: kinematics
AP (Application protocol) 201	Explicit draughting
AP (Application protocol) 202	Associative draughting
AP (Application protocol) 203	Configuration-controlled design

At the time of writing eleven further parts of STEP have reached the Draft International Standard status.

Parts of the STEP standard that have not yet reached the Draft International Standard (DIS) stage are available from the Solis information server at NIST (http://www.nist.gov/sc4/), subject to compliance with ISO copyright conditions.

Chapter 2

Constraints for Conceptual and Collaborative Design

Overview

Constraint Propagation Using an Active Semantic Network

Dieter Roller, Oliver Eck[1]

This section describes an approach for representing and propagating contraints in active database systems during product development. In this context, constraints are used for consistency checks of product models in databases as well as for the support of designers in product development. Concepts of active database systems are examined to represent and evaluate dependencies specified by designers during the product development process. Using concepts of object-oriented database systems, an approach for representing constraints by virtual objects is shown. A rule-based evaluation method for propagation of design modifications in the database is realized by Event-Condition-Action rules, known from the field of active database systems. This approach is illustrated by a realization of an active knowledge base, the Active Semantic Network.

Additionally, a method is presented to consider all dependencies originating during the product development process, especially if these dependencies cannot be formalized completely. Here, designers and experts are getting involved via automatic notification of the database system. The goal is to realize a constraint-based cooperation in product design. According to this approach, constraints can be propagated automatically by the system or interactively by a group of designers.

---

[1]Institut für Informatik, Universität Stuttgart, F.R.G.

Introduction

In the field of parametric and variational CAD, geometric constraints are mainly used to model mathematical relationships between numerical variables of model entities in the geometry of a design product [Rol90], [ShMä95], [Rol95a], [Rol95b]. Parametric features and geometric constraints led to the representation of parametric form features in solid modelling systems. Semantic knowledge about form features can be used to propagate modifications automatically.

Here, the goals are to make the design process more efficient and to reduce errors during design modifications. Figure 1 shows a gearing as an example of constrained features with geometric dependencies between form features in a parametric model. In this figure, according to [Rol95a], there is a geometric dependency between a key and a keyway and another one between a gear rim and an inner gearing.

Key   Keyway   Gear Rim   Inner Gearing

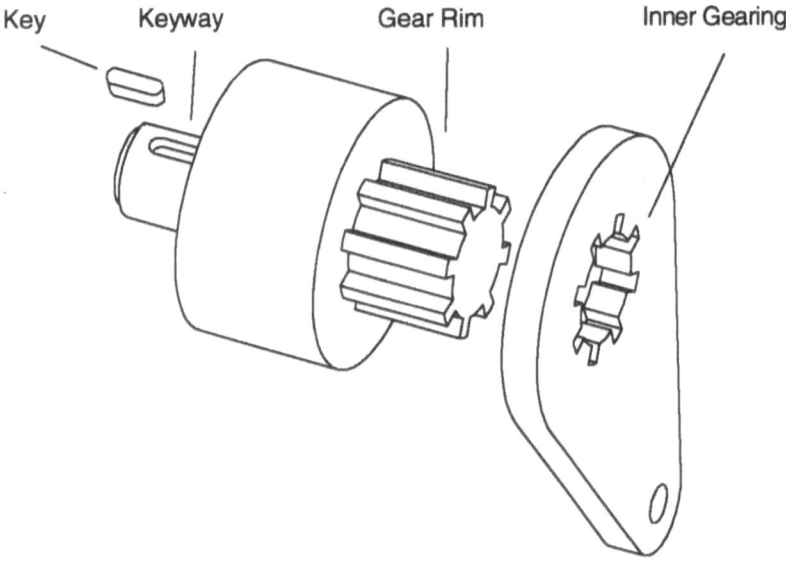

Fig. 1: Dependencies between different geometric form features

Modern product development is characterized by cooperative work of geographically distributed design teams. Although different designers are working on different subtasks, there are a lot of dependencies and logical relations between their design objects. In the approach presented here, dependencies among the requirements of a product and the properties of the product to be developed can also be modelled as constraints.

In this approach, all dependencies in an entire product model are regarded as constraints. This means that not only geometric dependencies are modelled as constraints but potentially also all other information relevant to product design. Particularly in the early phases of product design, where the most important decisions for successful products are made, constraints can be specified. A goal is to optimize critical factors like the design time, the costs and the quality of the product at the same time by visualizing the consequences of a design decision.

Modern product design is a process done by designers working interactively with computer aided design tools. In the approach presented here, constraint propagation does not play the role of an expert system which computes complex inferences isolated from the designers. Rather, the involvement of the responsible users is proposed to propagate constraints interactively if these constraints cannot be computed easily. In this way, possible errors in product design can be detected by the system and be satisfied in an interactive constraint evaluation.

In the following paragraphs, an approach to modelling of constraints in a shared design database is given and it is shown how to realize a rule-based evaluation method. This approach is based on an active, object-oriented database system. Object-oriented data representation seems to be most appropriate because of the complexity of typical product data. Object-oriented database systems are used more and more frequently for product models in product design environments but do not offer declarative mechanisms for explicit consistency constraints or active elements, like triggers [KoDi95]. Active database systems are database systems that allow users to specify actions to be taken automatically when certain conditions arise. In contrast to conventional passive database systems that only execute transactions when requested explicitly, active databases use rules to specifiy actions to be taken automatically when certain conditions arise [McC89], [Sto92].

Our approach to an active, object-oriented knowledge base is called Active Semantic Network (ASN). The ASN potentially represents all information relevant to product design and uses methods of active database systems and a rule-based evaluation method to propagate inferences. The goal of the ASN is to provide an engineering database that meets the database requirements for concurrent engineering in an active data-driven design.

In the following paragraphs, the fundamental concepts of the ASN are described. First, representation structures are presented to capture constraints in a network of data objects. Additionally, techniques of propagating constraints, concepts of transaction processing in constraint propagation and ways to involve experts are presented.

The ASN: An active knowledge base

The Active Semantic Network is a shared database system developed for supporting designers during product development and is realized as an active, object-oriented database system. The goal of the ASN is to represent all knowledge relevant to the product development process and to support geometrically distributed product design teams. Different approaches of modern database systems are adapted and integrated to meet the special requirement of cooperative product design. In this subsection, the basic approaches realized in the ASN are described, more detailed information can be found in [EcRo96] and [RBE97].

Product models are represented in the ASN by a network of nodes and links. The nodes of the network represent objects of the real world and the links relations among these objects. The structure of the network is specified in a meta model which describes the constructs available to represent knowledge in the ASN. The condition of a rule is given by the state of the network and the actions of the active component is able to propagate changes at one point of the network across the whole network or start external actions (e.g. at the graphical user interface).

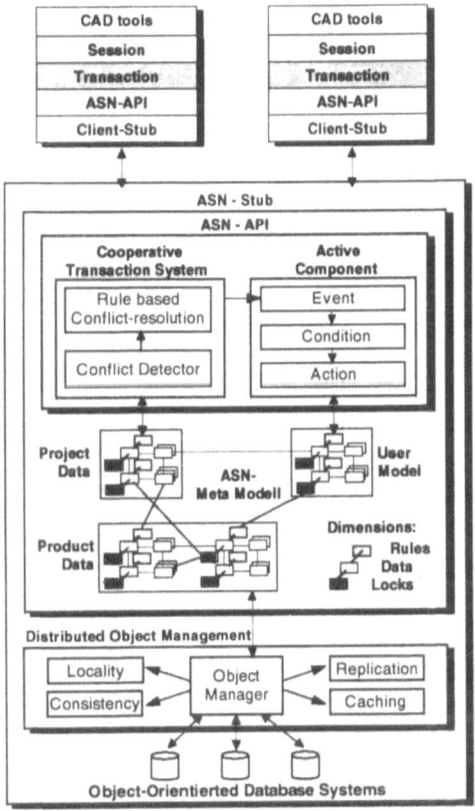

Fig. 2: Architecture of the Active Semantic Network

The active component of the ASN is able to propagate updates in the semantic network and to signal inconsistencies and conflicts to the responsible users. The ASN also supports teamwork and collaboration by initiating cooperative work when rules detect that the work of two or more designers does not coincide with eachother.

The goal of the active mechanism of the database system is not used to reason maintenance, to automatic revision of knowledge or to update product data unnoticed by their designers. Only the designers themselves are able to estimate the priority of different design decisions. Here, the task of the system is to support designers to visualize the consequences of a decision to his work and the work of his colleagues.

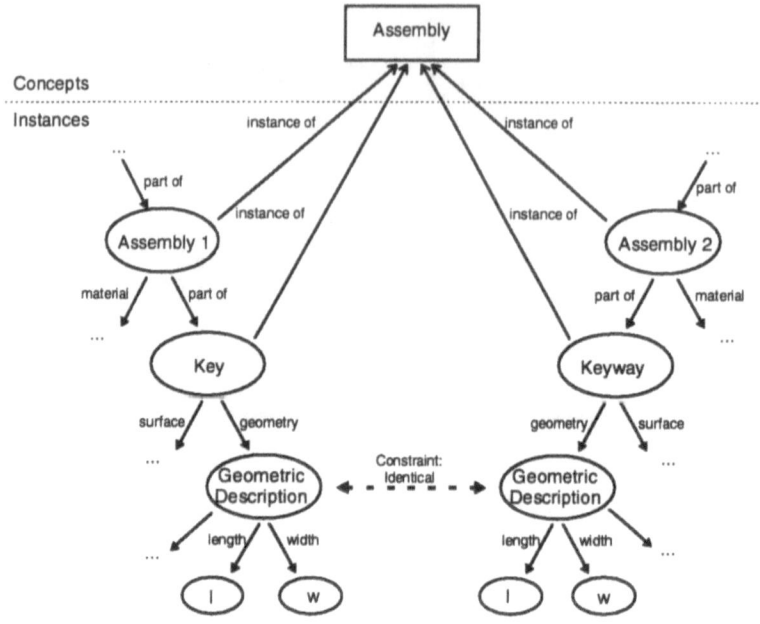

Fig. 3: Object-oriented product model of the gearing example

Similar objects can be described in the ASN by concepts. Analogous to the use of classes in object-oriented modelling methods, general objects are described and combined in concepts and common attributes of these concepts are described by slots. Concepts can be modelled in a subsumption taxonomy. In this taxonomy, slots of more specific concepts are inherited from more general concepts. Instances of a concept inherit all slots specified in this concept and concrete values can be associated with these slots. Complex objects can be modelled as "part of"-relations describing the components of objects.

The ASN uses mechanisms of distributed database systems to distribute design objects in a heterogeneous environment and to allow distributed access to shared data. Concepts of cooperative transaction models allow a group-oriented access to shared objects by multiple users [RBE97]. The active component of the ASN is used in the cooperative transaction system to localize and resolve locking conflicts between users. After an automatic notification, users are able to make an interactive conflict resolution and to modify objects cooperatively in the knowledge base using database transaction groups and tools for CSCW (computer supported cooperative work). Cooperative access to shared objects is realized by extended locking mechanisms which use knowledge about users and user groups in the knowledge base to support cooperative work. Objects in the database consist of three parts: the data itself, a set of associated rules, and locking objects for realizing the cooperative transaction system.

Figure 2 shows the basic architecture of the Active Semantic Network. The most important parts of this architecture are a meta model to specify the structure of the ASN, a database-independent programming interface, an active component for rule processing, a cooperative transaction model and a distributed object management.

Figure 3 shows a graphical presentation of a small part of an example product model representing the assembly of figure 1 in the ASN. In this notation, instances are presented as circles, concepts as rectangles and slots as arrows. A constraint between two objects containing exactly two variables and requiring dimensional identity of their values is indicated by an arrow between the related instances.

Data in the ASN is subject of constant changes and modifications during the product design process. Therefore, the structure of the network has to be flexible enough to process and support incremental increase of knowledge at run-time. But not only the data itself has to be flexible, but also the data structures and the set of rules, because all of this information can be modified during run-time. These modifications can be done by all users of the database system and therefore the database system has to maintain concurrency control mechanisms not only for data but also for data structures, rules and constraints. The active component has to adapt itself to a continuously changing set of rules. Inconsistencies, e.g. if a rule tries to access an object deleted by another user, are signaled to the corresponding users.

Representation of constraints in the ASN

The following paragraphs describe the representation of constraints within the ASN. Before a constraint can be propagated it has to be modelled. In this approach, this information is represented in the database and available to other designers to understand its meaning.

In [VSR92], it is shown how rules in an expert system can be used to describe geometric constraints in 2D geometric models. Rules are defined for geometric objects like triangles or parallelograms, etc., to evaluate the parameters inside these objects. Here, Event-Condition-Action (ECA) rules are used to realize the propagation of constraints. According to [McC89], ECA rules have the following attributes:

- Event: The event that triggers the rule
- Condition: A collection of queries that are evaluated when the rule is triggered
- Action: An action that is executed when the rule is triggered and its condition is satisfied

If all variables of a constraint are included in a complex object, this constraint can be modelled by defining one or more ECA-rules associated with this object. In the ASN, constraints are classified in the following way:

- Constraints in a single (complex) object
- Constraints between multiple objects

In the approach presented here, object-oriented methods are applied to constraints with the following advantages. Classes of geometric objects, e.g. circles or triangles, are collected together in concepts. Similar to this classification, general constraints can also be described as constraint classes. Using this approach, a constraint for parallelograms for example can be specified which checks the consistency of the geometry of all instances of the class parallelogram. In a concrete instance of the parallelogram constraint, additional conditions and actions can be added or overwritten but the most general specifications of all parallelograms can be reused in several parallelogram constraints. The pre-defined semantic of space and shape can also be modelled by constraint classes. These generalized models, including geometrical and topological constraints, can be reused in several design processes as basic definitions.

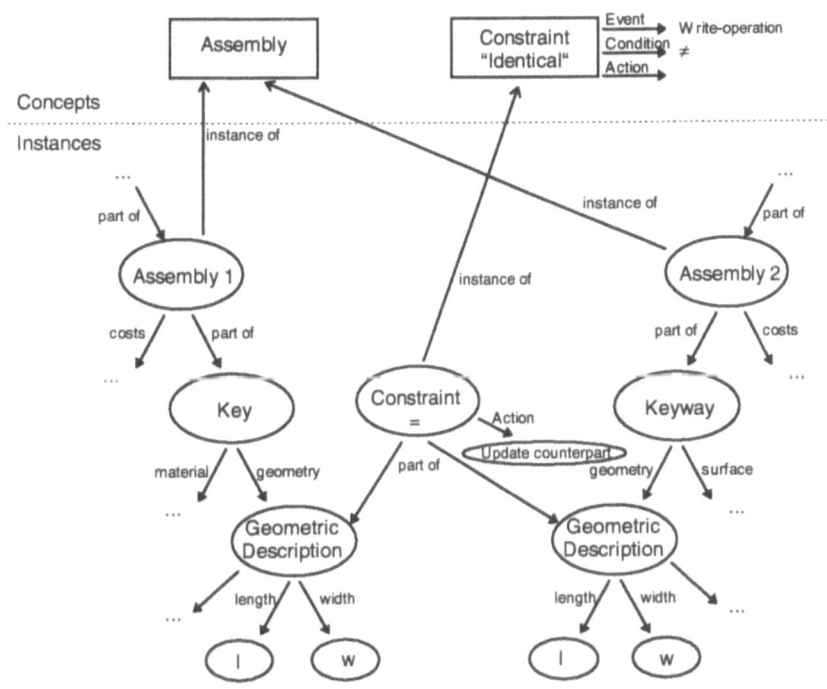

Fig. 4: Representation of constraints in the Active Semantic Network

Rules specified in [VSR92] can be classified as constraints inside single objects according to the above classification. In the example of figure 1, there is a dependency between a key and a keyway which is a simple constraint between different objects. More examples like this can be found in [Rol95a]. In order to model a constraint between different objects in the ASN, a new object is created which represents the constraint itself. Objects in the network which model no real item in the real world but properties like relationsships or constraints are called "virtual objects" in the ASN.

According to this distinction, the following types of concepts can be modelled in the Active Semantic Network:

- Object concepts: concepts representing real objects
- Constraint concepts: concepts representing constraints

Figure 4 shows the virtual object "Constraint =". The constraint concept "Identical" describes the identity of two geometry parts, and the instance of this concept models the fact that the instances "key" and "keyway" have the same geometry. The procedure of checking the dimensional identity of two geometries defined in the constraint concept can be reused for multiple geometries because this procedure is defined in a constraint concept which is inherited by all instances of this concept. The action of a rule describes the way to handle inconsistencies and may be different at different instances and so the way to propagate design modifications has to be remodelled in every instance. This example shows the advantages of the approach of constraint concepts. Procedures in the condition and action of rules can be reused in several constraints and may be overwritten if a special behaviour is needed.

Constraint propagation in the ASN

In the previous paragraphs, an object-oriented approach to represent dependencies was given. Next, an approach to constraint propagation using concepts of ECA rules of active database systems is presented.

As mentioned above, rule-based evaluation techniques can be used for constraint propagation. Here, it is shown how rule-based evaluation can be realized by ECA rules within active database systems. Using ECA rules, constraints can be model in the following way:

- Event: A write operation on one of the variables of a constraint
- Condition: A collection of queries that check if the constraint is satisfied
- Action: A collection of actions that are executed if the constraint is not satisfied in order to satisfy it

The complete specification of all dependencies among all variables of a constraint is only possible, if the constraint describes simple mathematical objects like geometric objects. Regarding dependencies between objects like product requirements and properties of the developed product, an automatic evaluation is not always possible any more. One reason is that these dependencies are too complex to model even in expert system. Another reason is that many of these dependencies are not specified by designers at the time the related objects are developed itself. Therefore, in the ASN, designers may get involved in an evaluation of constraints. In the case of errors during propagation, e.g. division by zero errors, or in the case of overdetermined constraints it is also possible to involve the responsible persons.

The ASN plays the role of an intelligent working platform, summarizing the design of all designers and supporting cooperative work on this data. Mechanisms to integrate knowledge-based information and communication systems with a knowledge base to support product development are used to involve the knowledge of human designers [EcRo96]. Here, models of communication infrastructure and the organisation of the company are necessary in the knowledge base to realize this approach. A consequence is that constraints can be used for automatic propagation when design data is modified as well as for the control of the actions of design teams. Automatic notifications in an active data-driven design can be used not only when certain conditions are not satisfied but also when conflicts between designers occur.

The effect of this approach for a designer is, for example, that when he makes object modifications which have complex dependencies on other parts of this or other objects, the constraint mechanism informs him about the consequences of his design modification and asks him to complete his modifications. If there are propagations to other objects for which other designers are responsible, these designers are informed and in this case constraints realize an adaptive workflow among designers. According to the constraint-based approach shown in [HEDL96], a "decision under constraints" function is realized here which allows negotiations based on constraints in cooperative design. Using the cooperative transaction model of the ASN, teamwork and cooperative access to shared data can be provided and constraints can be solved interactively between the responsible designers. In this way, a constraint-based cooperation in product design is realized.

The specification of coupling modes is very significant for the run-time rule processing of ECA rules. Rule events, conditions and actions of ECA rules are realized as database transactions and the coupling modes determine how these rules relate to database transactions. According to coupling modes, a different semantic of a rule can be specified. The definer of a rule has the flexibility to decide whether or not the conditions and actions of rules should be executed in the triggering transaction [DHW95].

E-C Mode	C-A Mode		
	immediate	deferred	decoupled
immediate	condition checked and action executed after event	condition checked after event, action executed at end of transaction	condition checked after event, action executed in separate transaction
deferred	not allowed	condition checked and action executed at end of transaction	condition checked at end of transaction, action executed in separate transaction
decoupled	condition checked and action executed in separate transaction	not allowed	condition checked in one separate transaction, action executed in a different separate transaction

Fig. 5: Coupling modes of active databases

In Figure 5, it is shown which combinations of coupling modes are allowed and which not. The coupling mode "immediate" indicates that the execution is done immediately inside the same transaction. "Deferred" means that the execution is done at the end of the current transaction and "decoupled" indicates that the execution is done in a separate transaction. In Figure 5, "E", "C" and "A" denote event, condition and action respectively.

The "immediate" and "deferred" coupling modes can be used in constraints to model dependencies which should be checked for consistency inside of the triggering transaction. Using these coupling modes, strong constraints can be modelled that prevent inconsistencies after user modifications by not allowing this modification. Using the "deferred" mode, inferences can be propagated across the network after manipulation. It is important here that the transaction lasts until the last inference is done and objects may be locked the whole time of calculation. The "decoupled" mode can be used if the condition or action of a constraint should be done in separate transaction. If a condition detects inconsistencies in an object using decoupled mode, the write operation is still finished and the original value of the triggering modification is not available any more. This coupling mode is especially appropriate for conditions and actions which may last a long time or include user interaction.

Summary

In this section, object-oriented and rule-based approaches are proposed to represent and propagate constraints in product databases. Concepts of rule-based evaluation methods are applied to active database concepts. The application of mechanisms of concurrency control to the representation of constraints was detected as a special requirement for the representation of constraints in cooperative database systems. In the ASN, constraints are modelled as regular database objects and subject to locking mechanisms. Using object-oriented concepts, general constraint information between abstract objects can be modelled as classes and instances of theses classes represent dependencies among concrete database objects.

Additionally, the integration of means of communication and cooperation was considered in order to solve constraints by designers when the constraints cannot be propagated automatically. Using an interactive constraint propagation method, constraints can be used not only for modelling of geometric constraints but also of other, more complex dependencies in product design, like the costs or quality of a developed product. Solving these constraints in product design, a cooperative constraint-based design process is realized. In order to propagate constraints automatically by the system without user interaction, mechanisms for applying ECA rules of active databases are examined for constraint propagation.

The ASN, including the presented active approach to constraint propagation, is realized in a first software prototype. The application of this prototype is evaluated by small amounts of product data. Next, an evaluation with a realistic amount of product data has to be done to obtain some practical experience of the presented approach. A problem of the presented system may be that the designers need to have modelling experience to model every detail correctly (e.g. the coupling modes of rules) in order to get the wanted active behaviour of the system. Here, a graphical user environment on top of the database system is needed that helps to visualize and manipulate objects and rules in the knowledge base.

The development and realization of the presented cooperative transaction model and the Active Semantic Network is supported in part by the Collaborative Research Center (Sonderforschungsbereich) SFB 374 "Rapid Prototyping" at the University of Stuttgart, Germany through the German DFG.

References

[DHW95]   U. Dayal, E. Hanson, J. Widom: Active Database Systems, in W. Kim (ed.): Modern Database Systems, ACM Press, pp. 434-456, 1995

[EcRo96]   O. Eck, D. Roller: Integration von wissensbasierten Informations- und Kommunikationssystemen zur Unterstützung der Produktentwicklung, in D. Ruland (ed.): CAD '96 - Verteilte und Intelligente CAD-Systeme, Kaiserslautern, pp. 242-255, 1996

[HEDL96]   M.J. Huguet, J. Erschler, G. De Terssac, N. Lompré: Negotiation Based on Constraints in Cooperation, in: Computer Supported Cooperative Work: The Journal of Collaborative Computing, No. 5, pp. 267-284, 1996

[KoDi95]   A. Kotz-Dittrich, K.R. Dittrich: Where Object-Oriented DBMSs Should Do Better: A Critique Based on Early Experiences, in W. Kim (ed.): Modern Database Systems, ACM Press, pp. 238-254, 1995

[McC89]   D.R. McCarthy, U. Dayal: The Architecture Of An Active Data Base Management System, ACM SIGMOD International Conference on Management of Data, pp. 215-224, 1989

[RBE97]   D. Roller, M. Bihler, O. Eck: ASN: Active, Distributed Knowledge Base for Rapid Prototyping, in: D. Roller (ed.): Proceedings of 30th ISATA, Volume „Rapid Prototyping in the Automotive Industries", Automotive Automation Ltd., Croydon, England, pp. 253-262, 1997

[Rol90]   D. Roller: Parametrische Formelemente als Basis für intelligentes CAD, in: K. Kansy, P. Wisskirchen (eds.): Graphik und KI, Informatik-Fachberichte 239, Springer-Verlag, pp. 92-102, 1990

[Rol95a]   D. Roller: Solid Modeling with Constrained Features, in: G. Farin, H. Hagen, H. Noltemeier (eds.): Geometric Modelling, Springer-Verlag, pp. 275-284, 1995

[Rol95b]   D. Roller: CAD. Effiziente Anpassungs- und Varianten-konstruktion, Springer-Verlag, 1995

[ShMä95]   J.J. Shah, M. Mäntylä: Parametric and Feature-based CAD/CAM, John Wiley & Sons, Inc., 1995

[Sto92]   M. Stonebraker: The Integration of Rule Systems and Database Systems, IEEE Transactions on Knowledge and Data Engineering, Vol. 4, No. 5, pp. 415-423, Oct. 1992

[VSR92]   A. Verroust, F. Schonek, D. Roller: A Rule Oriented Method for Parametrized Computer Aided Design, in: Computer Aided Design, Vol. 24, No. 10, Butterworth, pp. 531-540, 1992

Analyzing and Optimizing Constraint-Structures in Complex Parametric CAD Models

Reiner Anderl, Ralf Mendgen [1]

CAD-modelers have made a change from simple modelers of part geometry to sophisticated applications for representation and presentation of digital mock-ups (DMUs) of complex products. Paradigms like assembly modeling or design by features are common practice. Parametric modelers, that do not only model a product´s static structure but also try to keep the intent of a design by constraints or even derive the geometry of a part as a variant of a series of parts based on a constraint representation, are the state of the art. The designer who has to use these applications is confronted with an overwhelming functionality, giving him the possibility to represent almost all of his design decisions in the CAD model in different, more or less structured or comprehensible ways. Along with this progress in functionality the models created with these applications have become more and more complex. The development of functionality that supports the designer or even a team of designers and other users to understand, document, modify or reuse complex models has not kept track with the development of modeling functions. The presented work aims at providing this functionality by developing methods and tools for the analysis and optimization of structures of complex parametric CAD models. The state of the art of structuring and analyzing functionality within constrained CAD models is discussed and methods for the optimization of constraint-structures within CAD-models as well as for the analysis and comprehensible presentation of constraint-structures are introduced. Moreover a prototype system providing analysis functionality based on the proposed methods is presented.

---

[1] Department of Computer Integrated Design (DiK), Darmstadt University of Technology, Germany

Complexity of Parametric CAD-Models

Static (Explicit) and Dynamic (Parametric) Structure of CAD-Models

Today's CAD-systems offer many different functions to structure a complex model or mock-up of a product. In the following we will use the term model in the sense of a product's digital mock-up and not in the sense of a single CAD file. The components of this structure can be identified as assemblies, subassemblies, parts and features. The methods of structuring can be classified as static or dynamic.

The static structure in the sense of the product structure is defining structural components as aggregations of several other components. This is the case with assemblies that are defined as an ordered set of subassemblies and parts or with parts that are defined as an ordered set of features. The static structure in the sense of the model structure is defining additional component sets on different levels of detail that aim at improving modeling efficiency for both the user and the application. Files, layers, simplified representations or groups are examples of structuring methods for model structuring in addition to product structuring.

Things become even more complex if a parametric system is used because it adds a dynamic aspect to the structure of a model. We will not distinguish here between so called "parametric", "constraint based" or "variational" systems and use "parametric" as a generic term. In a parametric system, the presentation of a model to the user or the current shape of the model, that may e.g. be represented as a B-Rep structure (Boundary-Representation), becomes a current instance of a parametric representation of the model. This means that the parametric representation has the ability to generate different model instances depending on the current values of the parameters and on the constraints defined. This functionality ranges from simple modification of part's and feature's dimensions to modification of product structure. Examples are the amount, location or even presence of features in parts or of parts in assemblies that may be driven by parameters and constraints. Parameters may be defined for any of the structural components, and constraints may define relations between instances of any type of component. For detailed information about the different parametric methods and applications refer to [LiGo82], [Brue87], [ChSh90], [Roll95] or [AnMe96].

Figure 1 shows examples of CAD-models and table 1 shows some information about their structural complexity. The models represent

- a part with complex shape but without special parametrization (left),

- a gearbox parametrized for deriving a product series by torque and transmission ratio, but still with reduced complexity of shape (middle) and

- a shaft with integrated dimensioning calculations driven by forces, torques and bending moments [AnMe97] (right).

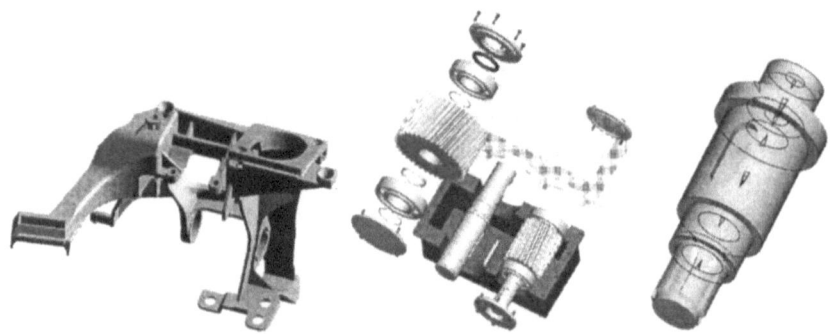

Fig.1: Examples of CAD-models

Structural Information	left	middle	right
Files	1	36	1
Layers/Groups	18/-	-/-	-/9
Features	368	665	80
Generations (Design History)	29	8	22
Geometric Parameters (Dimensions)	633	1363	1217
Geometric Constraints (between Features)	1207	1291	412
Non-Geometric Parameters	-	32	> 100
Non-Geometric Constraints	-	287	> 1000

Table1: Some information about the structural complexity of the models

Complexity will even increase, if parts have both, complex shape and complex parametrization. In the following we will use less complex exemplary models for the discussion of tools and the explanation of methods.

Problems Resulting from Model Complexity

Problems resulting from model complexity may be related to either the static or the dynamic structure of the model. In this paper, we will focus on the problems related to the dynamic structure of the model, comprising the detection or definition of parameters and constraints. Based on own practice in modeling, on experiences in teaching modeling and on contacts with industrial users, the following problem scenario could be identified.

When working with parametric systems designers are faced with some additional problems resulting from the use of parametric functionality. When using a system with constraint detection functionality (often implemented in so called "sketchers") the generation of shape may be fast and convenient, but the detected constraints (like paralellism or perpendicularity of lines) may be only correct for the current instance of the model and not for any of its intended variants that may be generated by parameter variation. Other problems result from constraint definition by the designer. Constraints such as equations or inequalities on dimensions or non-geometric parameters may be defined to derive or vary the shape of a model by calculations (affecting dimensions, structure and even presence of the model's components). Thus a dynamic model is created by defining a parametric (algorithmic) structure underlying the current instance of the model. Poor documentation and poor functionality of analysis and presentation of the model's parametric structure most often leads to problems of understanding or even to avoiding the use of the parametric functionality. In practice simply changing one dimension of a feature may cause changes throughout the whole model in consequence and affect parts or assemblies in different files (domino-effect). Therefore a user, particularly a user who did not create the model, is often not able to foresee the consequences of his actions.

Real life models using parametric functionality very rapidly become too complex to keep an overview of the evolving structure and the relations between the components of this structure. The effect is, that the creation of a parametric model often ends up in a model crash, meaning that the designer is no longer able to modify the existing model without causing changes he did not intend in consequence. Usually the designer has to redefine or delete large amounts of the model or even to restart modeling from scratch, either of which is very ineffective. These bad experiences often cause designers to reduce the amount of used modeling functions to keep the model simpler which, of course, leads to inefficiency too.

If it is difficult to create a model then it is even more difficult to reuse it for variation or modification purpose. The main problem is that there is no overall view of the model structure and therefore it is difficult to find a suitable starting point for modification without causing the domino effect. Specially this is the case if the model is reused after a longer period of time (the designer no longer has the structure of the model in mind) or by a different person than the model creator. Typical applications are variation of an existing product (and thus variation of an existing model) or reuse and adaption of part of an existing model in another product.

With distributed modeling or modeling in teams (Simultaneous Engineering, Concurrent Design) coming up more and more, the problem will increase, because several people will have to keep track with the evolving structure of a model that is created, modified or varied partly by any of them.

Resulting Requirements

It is obvious that the designer or a team of designers have to be provided with the right feedback of information from the model to deal with the problems described above. In order to achieve this, two main things have to be done. On the one hand methods have to be developed, which enable the designers to structure their models more comprehensively. On the other hand methods and tools for the analysis and optimization of constraint-structures in complex parametric CAD models have to be developed, to support the designers in analysing and structuring models.

Application and Analysis of Constraint-Structures in CAD-Systems

An Overview of Related Functionality in Commercial CAD-Systems

Most recent CAD-Systems provide functionality dealing with geometric and non-geometric parameters and constraints.

While geometric parameters are automatically defined when

creating geometry (e. g. in sketching mode), non geometric para-meters are defined by additional user input. The first just represent the dimensions of the geometry and the second may be compared with variables in a programming language, defined by name and type. Dependent on the CAD-System, non-geometric parameters may be defined within the scope of a feature, a part or an assembly (compar-able with the local or global scope of variables in programming languages), but always in the same way. There is not much more to say about the parameters, because the main functionality is dealing with the constraints.

Geometric constraints are defined when references between the components of the model´s static structure are defined. These constraints can be classified as

- constraints within features,
  (examples are parallel edges or coincident points)

- constraints between features, often called parent-child-relations,
  (examples are offset faces, parallel edges, extend to both sides of the sketching plane or extend until next face) and

- constraints between parts, also called placement-relations.
  (examples are mating faces or aligned axes; these constraints are more or less the same as those between features)

The first are defined or detected in sketching mode while the second may also be defined as attributes of the geometry generating function. The third are always defined by the user when creating assemblies. All these constraints are internally represented as equations but they may be expressed by names that classify the constraints (e.g. a parallel or mate constraint). Moreover all of them represent relations between elements of the B-Rep model (faces, edges, vertices, etc.), but may be related to the components of the model´s static structure for more convenient use.

Non-geometric constraints are always defined directly by user input. They are represented and expressed (presented) as equations or inequalities. Parameters used may be geometric or not. The number of constraints is not restricted but calculations are restricted in complexity, because control structures (like if-then-else, while, foreach, etc.) and more complex mathematical operators known from programming languages are supported poorly. Dependent on the CAD-System, non-geometric constraints may be defined for para-meters within features, between features or between parts. Regarding Windows®-based CAD-systems, a trend towards "outsourcing" non-geometric parameters and constraints to other applications like spreadsheet calculation programs can be observed. But this does not

directly lead to improvements of documentation and comprehensibility of constraint-structures. It is not obvious, which CAD
parameters are driven by external spreadsheets or which part of the
shape will change when spreadsheet parameters are modified.

Given all these different types of parameters and constraints and
looking at the amount of parameters and constraints in real life
models (table 1), it is obvious that very complex structures evolve. In
the following, we will discuss what functionality is provided to keep
track of the evolving structure.

Analysing Geometric Constraints within Features

The most common technique for creating geometry in feature based
parametric CAD-systems is sketching [PaWy85] [EHBE97]. Sketching tools are components of almost any recent CAD-system. These
tools do not only provide powerful functions for feature creation in a
parametrized model but also a restricted set of analyzing functionality for the constraints defined or detected in a feature. The main
problem concerning constraint analysis in sketching tools is the
local, single feature oriented scope. They only deal with constraints
within a current feature or between the current feature and its direct
parent features. Therefore, they are able to analyze variations of the
shape of the current feature if any constraint is removed or added, but
there is no preview of the changes that will occur in any of the children
of the current feature. The main reason is, that the development of
these tools aimed at user support in creating geometry and not in
analyzing models with constraints after creation. At the time of
model creation, an analysis of the current feature and its direct
parents may be sufficient, but an analysis of an entire model would
have to be performed feature by feature. This would be very time
consuming and would not provide a real overall view of the model´s
structure.

Figure 2 shows an example of a feature that is analyzed for constraints in the sketching tool of PTC´s CAD-system Pro/ENGINEER.
It is a rib (thick edges), that is designed to reinforce the housing of a
motor (thin edges). Detected or defined horizontal and vertical line
constraints, that result from aligning edges of the sketch with existing
edges of the part, are shown by the letters H and V near these edges.
The constraints may be disabled to see the effect on the current sketch.
The feature structure is reduced to all features in regeneration

sequence up to the current one. Any effects on children of the sketched feature are not taken into account when doing any modifications or redefinitions.

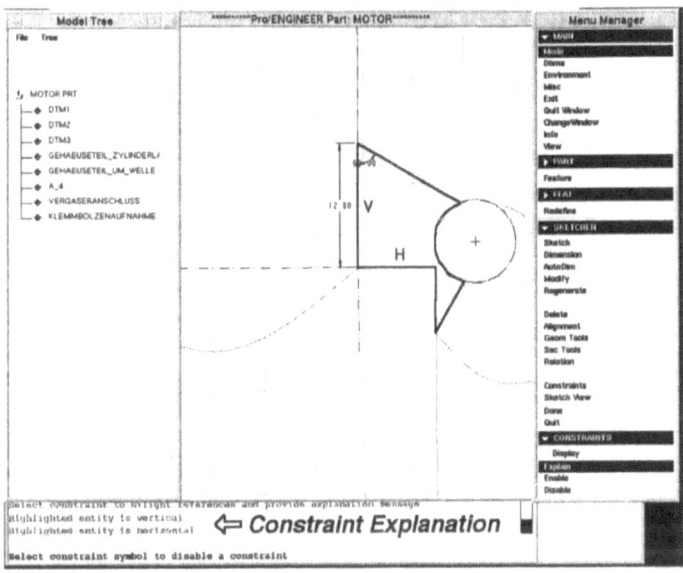

Fig. 2: Analyzing the geometric constraints in a sketch
(system: Pro/ENGINEER of PTC)

Analysing Geometric Constraints between Features or Parts

Things become more complex and interesting when we change from simple constraint sets within features to constraint sets between features of a part. As mentioned before, constraints between features may be created with any use of geometry generating functions or during any sketching process. These constraints are essential for the understanding of a parametric and feature based model, because they represent the most important information about the model's structure. Without them, a model would just be a collection of unrelated features.

Although these constraints are often called parent-child-relations, the structure of these relations does not represent a family tree, because any of the features may be related to any amount of other features throughout the model without regard of their generation in the modeling sequence. In this way, a structure is created, that may be classified as an acyclic and directed graph but not as a tree structure. In contrast to this, all analyzing functionality even the most recent

CAD-systems have to offer, is restricted to a query of direct parents or children of one single selected feature or a general view of the model's structure just presented as a list of features ordered in modeling sequence. These views of model structures are called model trees, part trees (even though there is no tree structure!) or more correctly feature managers. The idea behind these viewers was to have an alternative presentation of the model besides the geometric presentation. A modeler with a dual presentation interface enables the user to identify features in the presentation of his choice and then apply any of the feature-related functionality (modify, reroute, redefine, suppress, etc.). These viewers are powerful tools for doing this, but they are of minor use if the problem of model analysis is concerned. They do not provide an overall view of the model's constraint structure and may even make the user believe the structures are much simpler than they really are.

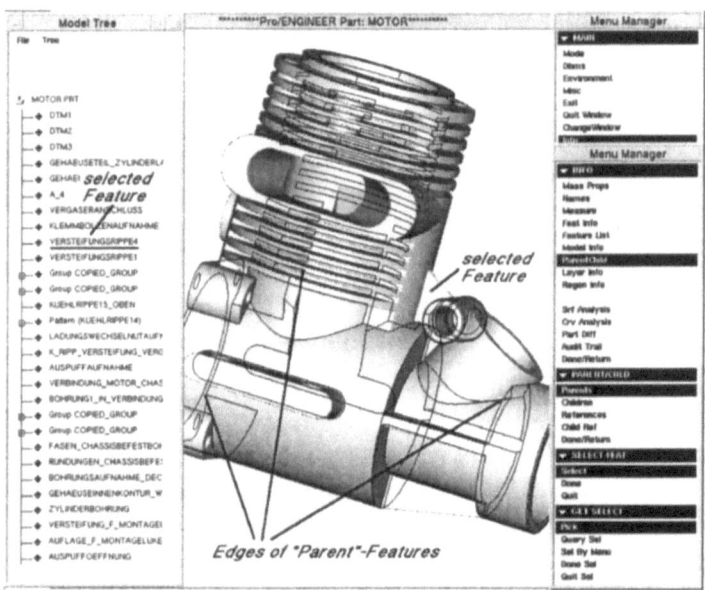

Fig. 3: Analyzing constraints between a selected feature and other features of a part using database queries (right side), as well as the sequence of features in a part using a feature manager (left side); (system: Pro/ENGINEER of PTC)

Figure 3 shows the result of a query for direct parents (edges in black) of a selected feature (the rib of fig.2) in a model as well as the model tree's structural view of the model. Only one generation of parent-child-relations is affected by the query and only a listing without any information about constraints or other structural information is provided by the model tree.

Analysing Non-Geometric Parameters and Constraints

Even though non-geometric parameters and constraints may already be defined within CAD-models for several years and this functionality has proven to be very effective for the derivation of variants or series of parts, for simple simulations as well as for integration of dimensioning calculations, there is no functionality for the analysis of non geometric constraint-structures up to now.

Non- geometric parameters and constraints may be related to assemblies, parts, features or specific dimensions of a model. Access to these elements is structured alike and restricted to lists of parameters and constraints of an assembly, part or feature. There is no direct information about which part or feature is affected by any of the constraints, which part or feature the driving or driven parameters of a constraint belong to or which constraints a parameter is used in throughout the model. If more than just a few parameters and non-geometric constraints are defined within a model, structures and the effects of modifications become so incomprehensible, that the overall benefit of this functionality becomes questionable. Without the provision of more effective analysis functionality, debugging of non-geometric constraint-structures while modeling or trying to understand those structures created by other designers, is often consuming more time than is saved by the application of this functionality.

Figure 4 shows an example of an assembly that uses non-geometric parameters and constraints. Based on some user input, relative position of parts, deformation of parts as well as several dimensions of features or even the presence of parts or features is calculated by non-geometric constraints. These parameters and constraints are spread all over the model and an overall view of their structure may only be obtained by reading the parameter and constraint listings of any of the affected features, parts or assemblies. The example shows the assembly of just a simple adjustable nozzle and a listing of the non-geometric parameters and constraints used for the calculation of the shape of the „Jet" (based on user input parameter values and the resulting deflection of some parts).

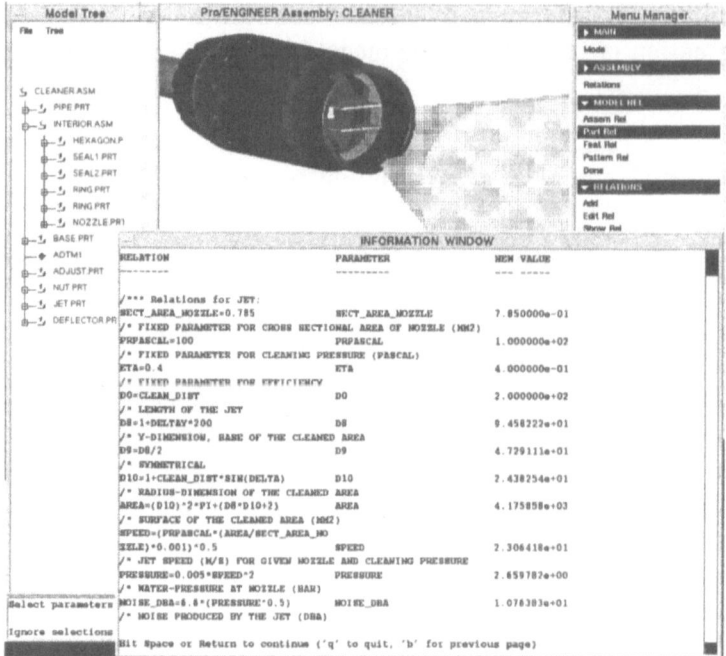

Fig. 4: Documentation of non-geometric constraints within an assembly
(system: Pro/ENGINEER of PTC)

Development of Methods and Tools for the Analysis of Constraint-Structures in CAD-Models

General Concept

To meet the requirements described above, methods and tools will have to be developed which deal with both, complexity as well as comprehensibility of model structures.

As complexity automatically increases with the amount of information within models and thus with the functionality of models, it may not always be reduced without undesired loss of information. Knowing this, the comprehensibility of models has to be improved by providing the designer with a presentation of the complete structural information of models in a more comprehensible way.

The desired complete information base for structural information of a model can be obtained by calls to the CAD system's database using the programming interface. In this way, all relevant information can be stored in a system independent database for structural information of feature based parametric CAD models, replacing hundreds of queries by a single analysis process. Standard data exchange interfaces may not yet be used for this purpose, because they only provide poorly structured non parametric information about models [IGES90] [AP203]. Standard data exchange interfaces with feature based [AP214] or parametric functionality [Chri96] [Prat97] are still subject of development or even research.

Provided with this database, new user interfaces can be developed that take full advantage of any available model information. For an effective and comprehensive presentation of model structures to the designer, the following methods are applied:

- different views on model structure in different user interfaces (views on parts or assemblies as well as on geometric or non-geometric constraints)

- dynamical and context sensitive presentation of information (parent-child-relations, non-geometric relations, conditions, or layering and grouping information)

- functions for the navigation through constraint-structures (browsing through the constraint net from any starting point)

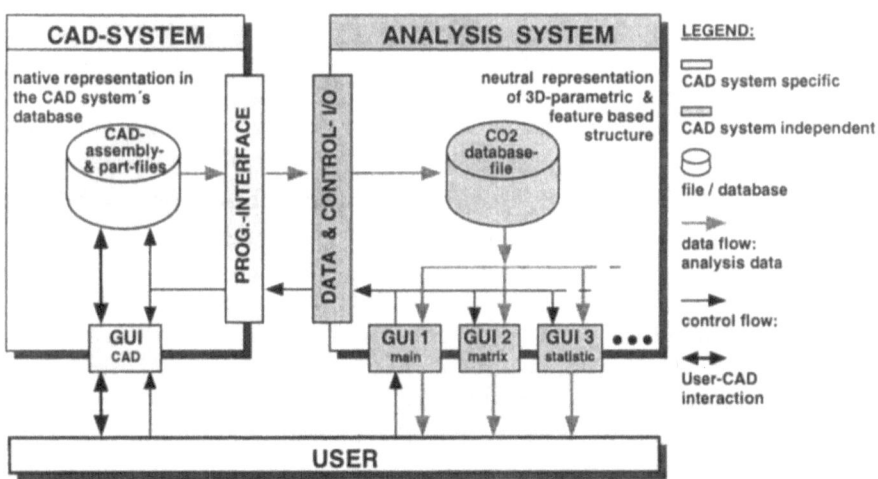

Fig. 5: General architecture of the tool

Figure 5 shows the general architecture of the analysis tool. A proto-type of this tool (called **Constraint Control**, or short: CO2) has been implemented with Tcl (Tool Command Language) [Oust94] and an exemplary interface to a CAD system has been realized for the Pro/-ENGINEER system of PTC [PTC97]. In the following, the concepts of the analysis functions will be discussed and the respective user interfaces will be presented.

Analysis of Geometric Constraint-Structures

As mentioned before, there is no analysis functionality available, providing an overall view of a model's geometric constraint structure (the parent-child relations between features). Since these relations between features of a model do not form a simple structure like a list or tree structure, alternative ways of presentation had to be developed. Knowing that a feature may be related with any amount of prior features in the modeling sequence by an "is child of" relation and with any amount of features later in the modeling sequence by an "is parent of" relation, these parent-child relations may be presented in form of a n*n matrix where the columns as well as the rows represent the model's n features.

The matrix is read row by row, showing the parents of any feature in the lower triangle and the children of any feature in the upper triangle. The features are presented by their ID-number and an icon showing the type of feature (swept profile 🖼, round 🖼, rib 🖼, etc.). Any other information about any couple of features in a parent-child relation is presented dynamically at the bottom of the matrix window.

For example in figure 6, feature ID 335 (the rib mentioned before) has datum planes ID1 and ID5 as well as swept profile ID 7 (cylinder), rotated profile ID 93 (crankshaft housing) and swept profile ID 316 (housing for a bolt) as its parents but no feature as its child. Information about the row's and the column's feature is dynamically bound to the cursors presence in any of the cells. In this way information can be obtained about the feature's name, whether it is a copy of another feature or whether it is assigned to any group or layer within the model. Moreover there is an associated and dynamic display of this feature information in the CAD-window. The feature of current interest as well as its parent- or child features are shaded in different defined colors and the features outside of the current scope are shown transparent. This is a very effective way of getting an overview of feature dependencies, using a structural as well as an associated geometric view of the CAD model.

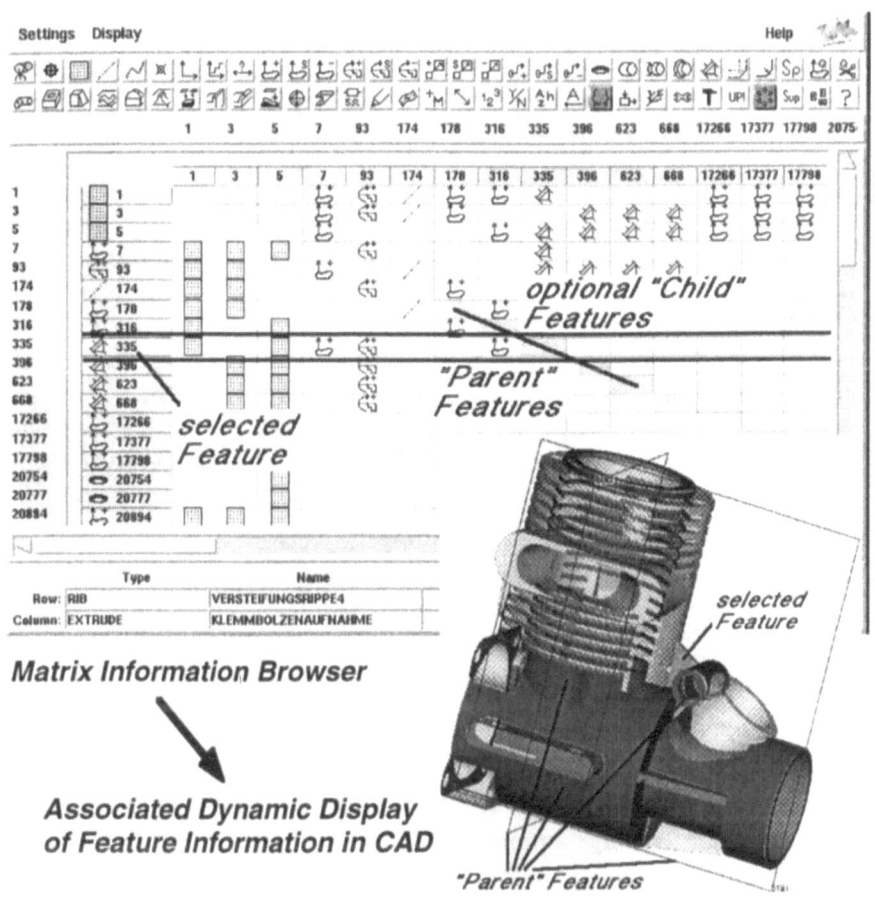

Fig. 6: Viewing and browsing geometric constraints between the features of a part

Analysis of Non-Geometric Constraint-Structures

The non-geometric constraint structure of a model is characterized by the non-geometric parameters and the non-geometric constraints that relate these parameters with each other or with the dimensions of model components that are to be varied as the effect of these parametric calculations. Parameter values are input data or calculated by the constraints. Constraints are relations between parameters or restrictions on parameter values. There are two types of relations, equations and inequalities. When working with a pure parametric system each equation or inequality corresponds with exactly one parameter that is defined or restricted by this constraint in explicit form and is called the driven parameter (parameter c < = > f(parameters a,b)). This form of directed relations is the standard in most CAD-systems. The other parameters of the equation or inequality are called the driving parameters. Some CAD systems even allow the definition of control structures (like If-Then-Else) that are used to define conditions under which some constraints may be applied or not.

Given this, we can present the non-geometric constraint structure of a part as a list of parameters in the sequence of calculation from input parameters to the driven dimensions of the geometry. The parameters are primarily characterized by the type that is shown by an icon (number 🔢, character 🔤 or boolean 🔳 input or driven 🔻) and the source (assembly or part file) that is put in front of the parameter name. Based on this presentation any other information related with the non-geometric constraint structure can be obtained interactively by selecting parameters. Complete information related to a parameter consists of:

- the defining relation (that drives the parameter) or alternatively the definition of the input type,
- any restrictions on the parameter value or on the application of the whole defining constraint (resulting from control structures),
- any other relations the current parameter is used in (is driving),
- the source of these relations or, if the driven parameter is a dimension, the feature where this dimension is used to define geometry,
- and last not least any comment of the programmer that is available for the explanation of the constraint.

Figure 7 shows an interface to parameters and relations in non-geometric constraint-structures that meets all the requirements discussed above. The example data are from the model of a cleaner also shown in figure 4. The left shows the list of parameters of the whole parametrized assembly. The parameter DeltaY_J of the part jet is selected and the relations it is defined by or driving are displayed. The selected parameter is an input value for the part-model of the jet, that is assigned a value by a call from the assembly cleaner where this part is used in. In this case, the input value for the deflection in Y-direction is derived by the assembly parameter twist-angle. In addition it is shown that the selected parameter is used to drive the parameter D8, a dimension of feature ID7 of part jet, in the next constraint. The figure only shows directed constraints because the model has been created with a CAD-system that is restricted to directed constraints. Up to now, the same restriction applies to the tool, because it has only been used together with CAD-systems like that.

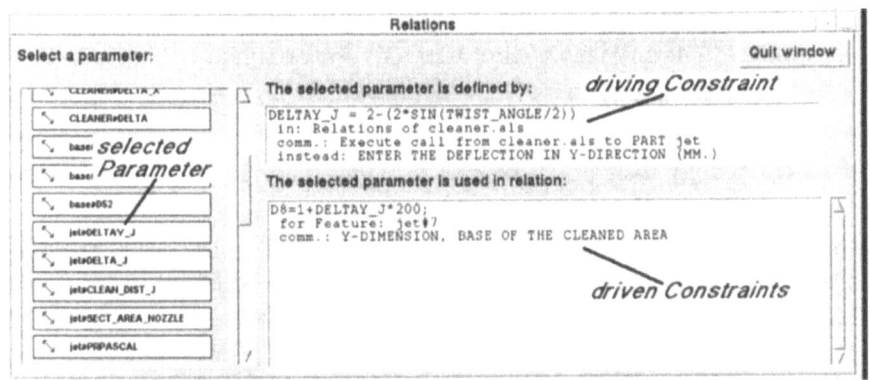

Fig. 7: Analyzing non-geometric constraint-structures within a part or assembly

Navigating Through the Constraint Structure

With growing complexity of models, navigation through the constraint-structures becomes more and more important. First of all this is the case with assemblies. Combining the views of the geometric and the non-geometric constraint structure,

- the sequences of parameters and features of each part-file of the assembly as well as

- the sequence of parameters and parts (features, subassemblies) of each assembly file

have to be presented to the user. One possible realization for a GUI is a chart, that is structured in columns. Any file (part- or an assembly-file) is represented by a column showing the geometric elements of that file (the subassemblies and parts of an assembly or the features of a part) and an additional optional column showing the geometric and non-geometric parameters of that file that are driven by constraints. Within any row, the geometric elements or parameters are represented in sequence of calculation (sometimes also called regeneration in CAD-systems).

Any geometric element or parameter is assigned an icon for type (e.g. rotated profile ⬛, numerical input ⬛, calculated parameter ⬛, etc.) and a name or identification number. All available information about any element in display or about the relations between the elements in display is accessed interactively.

Depending on the type of element, information about

- the parents or children of the currently selected feature,
- the parameters driving or driven by the currently selected parameter, and
- the conditions that apply to the currently selected feature or parameter

is accessed and presented as arrows in different colors from element to element. Any element may be selected as a starting point for a navigation through the whole structure, because repeated selection of an element causes the display e.g. of parents, grandparents, etc. (children, grandchildren, etc.; parameter sequences from inputs to dimensions) through the whole model.

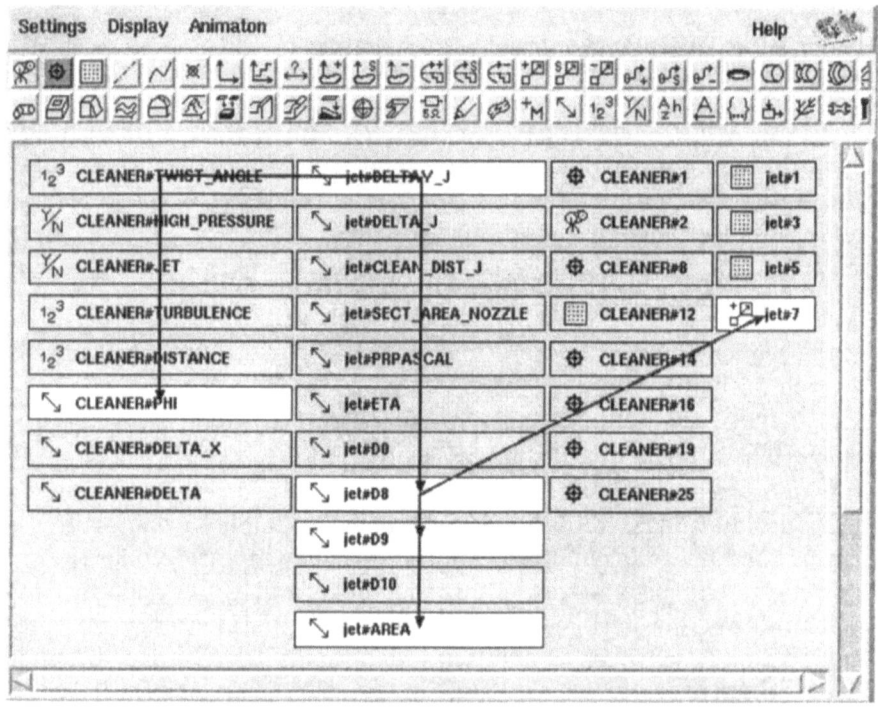

Fig. 8: Navigating through constraint-structures of an assembly

Figure 8 shows part of the browser interface for the assembly of the cleaner. In the example, the non-geometric constraint structure is browsed from an input parameter of the assembly file (Cleaner#Twist_Angle), to the dimension D8 of feature ID 7 in the part model of the jet. In the first step of calculation, parameter Phi of assembly cleaner is derived from the input value and input values of other parts are assigned (e. g. DeltaY_J of part jet). In the second step, the dimension D8 is derived from DeltaY_J within part jet. In the third step, D8 is used to derive other parameters and used as dimension in Feature ID 7. In addition, information windows showing any available information for any of the geometric elements or parameters may be displayed on selection.

Methods for the Optimization of Constraint-Structures in CAD-Models

Possible strategies for the optimization of constraint-structures in CAD models are diverse and depending on the objective of optimization. Objectives may be the fast creation of models, fast parametric calculation of a model's shape or comprehensive structure to simplify the understanding of the model. The last one will be the matter of the following discussion. As mentioned before, structural complexity may not be avoided in general but may be reduced by applying structuring methods or managed by applying analysis tools.

Since designing and creating CAD-models is a very complex engineering task, where most of the reasons for any detailed design decision are not represented in computer interpretable form, it is obvious that automation of the model structuring process (e.g. by algorithms) is only second choice. The development of methods for structuring models comprehensively, as a set of rules for the designer to be taken into account during the modeling process, is a much more effective solution for the problem of structural complexity. The two ways of optimization may be characterized as follows.

- Avoid unnecessary complexity of structure
  Given a set of modeling rules, the designer is able to reduce structural complexity based on his engineering knowledge during the modeling process.

- Rearrange structure to reduce complexity
  A restricted set of modeling rules may be represented in a computer interpretable way, thus enabling an interactive optimization process based on an existing model.

Optimization Strategy

Taking into account all the possibilities of defining and analysing structures of models discussed above, it is obvious that the complexity of these structures increases with the number of elements in the model (parameters, features, parts, assemblies) and the number of relations (geometric and non-geometric constraints) between them. Better comprehensibility may be achieved by

- structuring the whole model into well defined regions of interest (aggregation of elements),
- reduction of the amount of individual or relevant model elements (typing of elements),
- and last not least a structured approach of creating constraints.

Better comprehensibility from the beginning of the modeling proces on directly leads to a better understanding of the evolving model structure. This results in a very determined straight forward style of modeling, reducing try-and-error effects and thus in a reduction of created features, parameters and constraints.

Major improvements can be achieved by trying to modify the unstructured network of relations within a model towards a tree structure. In most models, some invariant features that are not affected by any modification or variation of the model can be identified and used as roots of the structure. Moreover, some kind of branching of the structure can be done by identifying regions of interest. A region of interest may be an aggregation of features that represent part of the shape (e. g. motor-block, cylinder-head or connection to the carburettor in the model of the motor) or that are related to a function of the model (e.g. cooling or transmission of forces in the model of the motor). It is typical, that features within one of these regions of interest have many relations between each other, but are only loosely connected to features outside. Within CAD systems, grouping and layering are commonly available functions for the aggregation of elements.

Typing of features leads to a drastic reduction of the amount of features that have to be taken into account when analysing geometric constraint-structures. In a properly structured model there are many features like rounds, chamfers, drafts, etc. that just represent details of the geometry and should not be used for parent-of-relations to any other features or parts. These are just hundreds of leaves in the structure that do not play any important role in the structure and may be classified as secondary or not relevant. Moreover many features or even sequences of features may be modeled as copies of existing ones, because their dimensioning scheme and reference scheme is identical and not only because of identical shape. These features or sequences of features represent identical partial structures of the model. Feature based CAD systems support typing of features as well as copying functionality that comprises definition of simply copied, mirrored or patterned sequences of features.

Improvements of the geometric constraint structure may be achieved by applying startegies of constraining. One of the most common strategies is to reduce the scope of other features that may be used when creating constraints between the current feature and the ones before. For example, features within a branch, features of a special type (datum features/ geometry features) or invariant features (sometimes a skeleton of axes and planes) are to be preferred when creating constraints. Improvements of the non-geometric constraint structure may be achieved by taking into account methods of structured programming. Using non-geometric constraints within CAD models and first of all assemblies is much the same as programming. Parts within a parametrized assembly may be compared with procedures that should be called with a well defined set of parameters and that return the derived shape of the part. Since this is not directly supported by the user interfaces of the CAD systems, most times relations are just spread all over the model.

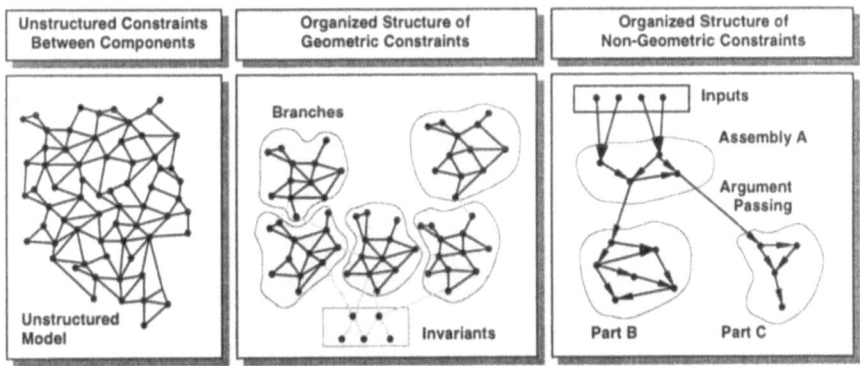

Fig. 9: Organizing constraint-structures comprehensively

Figure 9 presents the most important of the structural effects discussed above in graphical form. An example showing the effects of an organized geometric constraint structure will be given in figure. An example of an organized non-geometric constraint structure has already been presented in figure 8.

Avoiding Unnecessary Complexity of Structure

This approach is always feasible, if a new model has to be created. It uses a set of modeling rules for parametric modelers based on the optimization strategies described above. Such rules have been developed at DiK [Geis96] [Stur97] and have been testet on exemplary CAD-models from industry [Cras97].

Figure 10 shows that optimization based on these rules really leads to reduction of complexity as described above. The exemplary models are a casted housing of a gearbox (created with solid features) and a sheetmetal part (created with surface features). The results in the table show, that the amount of features as well as of parameters and constraints has been reduced drastically. Moreover it has been proved that the reduction of features does not simply lead to the creation of fewer but more complex features. The number of parameters or constraints per feature did only increase a little, because many datum features (planes. axes, curves) with only few parameters and constraints could be removed.

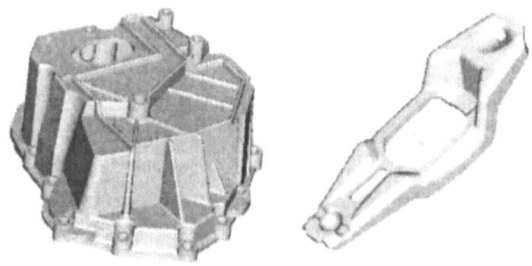

	orig.	new	diff.	orig.	new.	diff.
Sum Features	442	244	-49%	140	79	-43%
Sum Geom. Par.	1193	675	-43%	163	133	-18%
Sum Geom. Cons.	1807	1166	-35%	607	239	-60%
Geom. Par./Feat.	2.69	2.76	+2.5%	1.16	1.68	+45%
Geom. Cons./Feat.	8.18	9.55	+17%	8.66	6.04	-30%

Fig. 10: Effect of organizing geometric constraint-structures

Rearranging Structure to Reduce Complexity

This approach is especially useful, if an already existing model has to be optimized. Based on the modeling strategies and conventions discussed above, an automatic check of the model structure may be performed. Violations may be detected and restructuring may be performed in interaction with the user. This works well, if strategies for constraining (e.g. "always create references to datum features", or "do not create references to secondary features like rounds") are to be checked. Checking algorithms like that are used in the presented tool as well as in commercial checking tools for CAD-systems.

The next step in development will adress user support in structuring models by identifying features for copying, for grouping and reuse within the model, unnecessary datum features, invariant features or structural branches of the model (these are only exemplary points of interest). The final goal is to develop an optimization assistant that may as well be used during modeling sessions, performing checks and suggesting structural optimizations after every significant modeling step (feature creation or modification).

Acknowledgements

The authors want to thank all the students and faculty staff involved in this work for fruitful discussions about modeling strategies and model complexity as well as for the provision and analysis of example data and the implementation of the prototype. Last not least thanks to Robert Bosch GmbH and Audi AG for the provision of "real life" CAD-models to test and improve the methods and tools.

References

[AnMe96]   Anderl, R; Mendgen, R.: Modeling with constraints - theoretical foundation and application; in Computer Aided Design, Vol. 28 Number 3; Elsevier Science Limited - Butterworth Heinemann, Oxford, 1996

[AnMe97] Anderl, R; Mendgen, R.: Konstruktionselemente in
          parametrischen 3D-CAD Systemen; in VDI-Berichte 293 -
          Features verbessern die Produktentwicklung; VDI
          Verlag, Düsseldorf, 1997 *(in german)*

[AP203]   N. N.: ISO 10303 Product data representation and
          exchange, Application Protocol 203: Configuration
          controlled design; International Organisation for
          Standardisation, Geneva

[AP214]   N. N.: ISO 10303 Product data representation and
          exchange; Application Protocol 214: Core data for
          automotive mechanical design processes; International
          Organisation for Standardisation, Geneva

[Brue87] Bruederlin, B. D.: Rule Based Geometric Modeling;
          Dissertation submitted to the Swiss Federal Institute of
          Technology, Zürich; Diss. ETH No 8382, 1987

[Chri96] Christensen, Noel C.: Parametrics framework proposal;
          Working draft of the ISO/TC184/SC4/WG2 working group
          on the introduction of parametrics into STEP; available
          via   http://www.nist.gov/pub/subject/sc4/paramet/short/
          framewk/dec96/, 1996

[ChSh90] Chung, J. C. H.; Schussel, M. D.: Technical evaluation of
          variational and parametric design; Proceedings of the
          ASME computer in engineering conference; Boston,
          Massachusetts, 1990

[Cras97] Crass, M.: Konzeption und Überprüfung von Methoden
          zur Strukturierung und Dokumentation des Modellier-
          ungsprozesses in CAD-Systemen; Diplomarbeit, DiK,
          TU-Darmstadt, 1997 *(in german)*

[EHBE97] Eggli, L.; Hsu, C. Y.; Brüderlin, B. D.; Elber, G.:
          Inferring 3D models from freehand sketches and
          constraints; in Computer Aided Design, Vol. 29 Number
          2; Elsevier Science Limited - Butterworth Heinemann,
          Oxford, 1997

[Geis96] Geissler, H.: Konzeption von Methoden zur Struk-
          turierung von Volumenmodellen unter dem Gesichts-
          punkt der Änderbarkeit und Wiederverwendbarkeit;
          Studienarbeit, DiK, TU-Darmstadt, 1996 *(in german)*

[IGES90] N. N.: Initial Graphics Exchange Specification, Version
          5.0; NIST, Gaithersburg, 1990

[LiGo82] Light, R; Gossard, D.: Modification of geometric models through variational geometry; in Computer Aided Design, Vol. 14 Number 4; Elsevier Science Limited - Butterworth Heinemann, Oxford, 1982

[Oust94] Ousterhout, John K.: Tcl and the Tk Toolkit; Addison-Wesley Publishing Company, Reading, Massachusetts, 1994

[PaWy85] Pavlidis, T.; van Wyk, C.: An automatic beautifier for drawings and illustrations; ACM Computer Graphics, Vol. 19, 1985

[Prat97] Pratt, Michael J.:Extension of STEP for the Representation of Parametric and Variational Models; in Roller/Brunet (publishers) "CAD-Systems Development - Tools and Methods", Springer, 1997

[PTC97] N. N.: The Pro/ENGINEER CAD/CAM/CAE system - Version 18 manuals; Parametric Technology Corporation, Waltham Massachusetts, 1997

[Roll95] Roller, D.: CAD - Effiziente Anpassung und Variantenkonstruktion; Springer Verlag, Berlin Heidelberg New York, 1995 *(in german)*

[Stur97] Sturm, D.: Erhöhung der CAD-Datenqualität durch transparente Modellstrukturierung und Abbildung von Konstruktionslogik; Diplomarbeit, DiK, TU-Darmstadt, 1997 *(in german)*

Design Support Using Constraint-Driven Design Spaces

Attila Csabai[1], Joachim G. Taiber[2], Paul C. Xirouchakis[1]

During the conceptual phase of product design, rapid generation of the rough assembly description and capture of the product layout should be supported. This paper introduces a top-down approach for creating the product layout and facilitating subsequent detailed design. For this purpose "Design Spaces" are introduced. These design spaces represent the conceptual boundary of the components as well as references for defining relationships between the functional parts of the assembly. These functional units can be connected together by means of interface features and constraints. Once the design spaces and the relationships have been set up correctly the task of refining the components geometrically can be distributed and done in parallel. This concurrent work is controlled with access rights and well-defined responsibility areas. Since the layout and the component level are strictly delimited, kinematic analysis as well as layout modifications can be done at any stage of the product design without any explicit information about the geometry of the components. This paper deals with the methods used to set-up and analyse the product units using kinematic constraints.

[1] LICP, École Polytechnique Fédérale de Lausanne, CH

[2] Precisionsoft AG., CH

Introduction

The presented top-down method, called Layout Design Tool (LDT) facilitates capturing the human intention of the constructional design by providing tools and methods for doing this in a natural manner. This means that the design process starts with the overall definition of the functional behaviour of the product and the detailed description of the components remains for the design refinement stages ("minimum commitment modelling" [GuSt96]). Since there must be well defined connections between the components, the project can be separated into different tasks. It is possible that a task can also be further divided, which indicates the hierarchical nature of the method.

In the layout design abstract entities are to be used to form the layout. The geometry of these abstract entities should not be too complex because of the need for good overall view, fast simulation, easy changeability [LaFe96]. Although, several works use simplified solid models of the component to be shaped [LaFe96], the LDT applies "containers" called design spaces which have, nevertheless, geometric properties, but are represented in a different way with respect to the detailed components. The idea of the object-oriented design space approach can also be found in [ScSc96].

To summarise the key features of the presented approach could be:

- The design process can be started without any information about the detailed geometry.
- The LDT can be an efficient tool in order to create different product variants and to trace back the design modifications.
- Early stage kinematic analysis can be done without any information about the detailed geometry.
- The division of the product into conceptual units fits fairly with the modular design approach.

The LDT is intended to be a part of a larger system. For this reason several operations, such as geometric refinement or constraint solving, can be done outside the LDT environment.

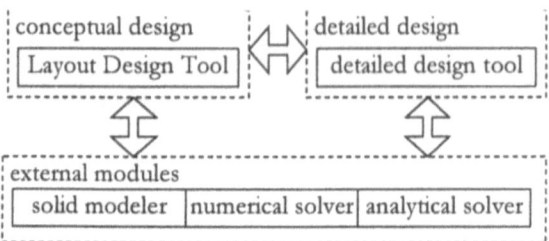

Figure 1. Connection of the LDT with other modules

The geometric refinement of the components can be accomplished by a solid modeller, the layout elements can be arranged by means of an analytical constraint solver module. The kinematic analysis of the non-rigid layout is rendered possible by a numerical constraint solver. Therefore, the product model used by the LDT and the internal representation of the external modules must be convertible into each other.

Description entities (terminology)

The initial, abstract representation of the product is called *product layout*. The layout principally defines the structure and the arrangement of its functional components as well as the way they interact each other. Once the layout has been defined, an early stage kinematic analysis can be performed, while the *components* of the assembly can be further refined and optimised.

The units of the product layout are the so-called *design spaces*, which are assigned to design engineers individually. The design spaces are characterised by their shapes and parameters. Each design space can be given a read/write access attribute, enabling only selected persons to view or modify it.

Interface features are sets of geometric entities like point, vector, plane and are used as datum references for the *constraints* and for the *connection form features*. The interface features are given relative to the design spaces, therefore, they are connected to the design space and not to the component inside. Several works use markers on solid bodies [Kram92][SeMe96], or assembly definition frames [Haug89]. The purpose of the interface features compared to these is that they are the same from the point of view of the constraint definition, but they also stand for references for the connection form features to be designed at the geometric refinement stage.

Constraints are used to establish relationships between the design spaces. They can be defined for two interface features, this way the necessary geometrical information will be obtained from the latter ones.

The *component* is a functionally well defined part of the assembly. This can be either a single part or a sub-assembly. This is quite similar to other interpretations [SeMe96], but does not matter whether the component is rigid or not; this is the responsibility of the unit designer. The main restriction is that the component must be adaptable for the design spaces and the interface features.

Connection form features are used to physically realise the intended connection, therefore, they should not depend on the components to be modified with them. For this purpose they are attached directly to the interface features to insure that their position and orientation reflect the conception of the layout designer. Nevertheless, when a component is to be exported from the LDT environment it should be possible to "freeze" the connection form features into the component in order to conserve the modified shape.

The relationship among the description entities is shown in Figure 2.

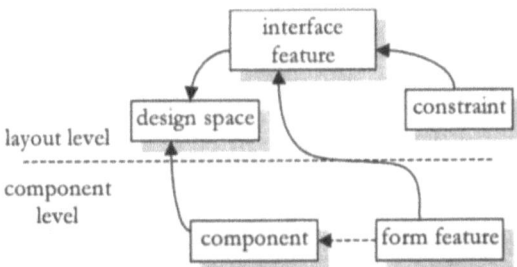

Figure 2. Relations between the layout elements (the meaning of the arrows: "with respect to")

Layout hierarchy

The designer can use the layout module in order to specify the product structure in a top-down manner, that is, i) roughly define the geometry of the components by introducing abstract entities, ii) determine the relationships between those abstract entities, iii) arrange the layout in space by using design spaces and interface features, and iv) verify the rigidity of the layout (perform a kinematic analysis, if it is possible or necessary). After this work the geometric refinement of the components can be started.

The product to be created is a set of functional units which occupy a well defined area of the real space, that is the *model space*. The design spaces are parts of the model space occupied by one component which can be either a sub-assembly or a single part. The main point is that it must be separated from the other parts functionally. The assembly hierarchy is in the form of a part-of tree *(assembly hierarchy tree)* where the root is the product, the interior nodes are the sub-assemblies, the leaves stand for the parts of the product while the edges mean the part-of relations between the sub-structures. In the *product layout tree* the *model space* is the root and the *design spaces* can be either the nodes or the leaves (Figure 3.)

There are *unit designers* assigned to each design space who are responsible for the contents of the design space. There is a "main designer", the *layout designer* who is responsible for the product layout; his main task is to organise and maintain the connections among the design spaces. If a unit designer is working on a sub-assembly, his design space can be further divided; in this case he functions as a layout designer of the lower level design spaces as well.

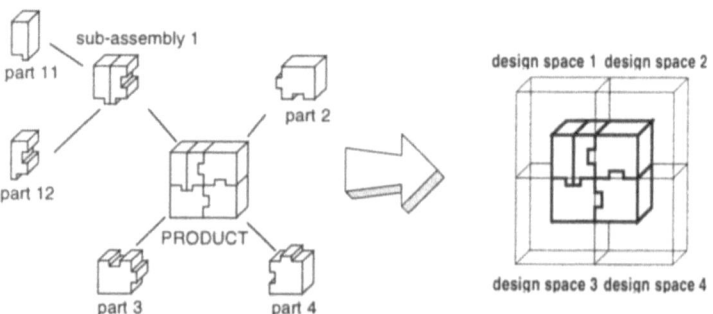

Figure 3. The assembly and the product layout hierarchy

Design spaces

The contents of a design space may consist of both parts and sub-assemblies at the same time. In the latter case the unit designer can decide if it is worth subdividing his design space into lower-level design spaces or whether he treats the sub-assembly simply as a compound component. The disadvantage of the latter case is that the LDT cannot treat the inner sub-assemblies as a lower level layout.

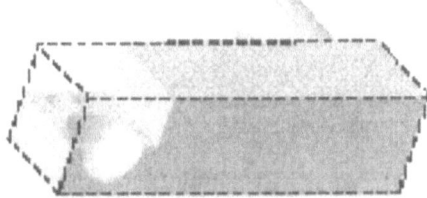

Figure 4. Example for design space. Note that the pin connection form feature for realising a connection is outside the design space.

The design spaces can be defined by determining the area of the real space they occupy. They should not have too complex geometry, because the goal is to create and handle them rapidly. However, on the other hand their shape should be sophisticated enough to reflect the component inside. According to the current approach, design spaces are defined as polyhedral objects. There are also constraints (parallel, perpendicular, offset, etc.) specified for the boundary elements (faces, edges) of the design space in order to make the parameterised changes possible. The designer can define a unique shape for his design space or he can choose from the library of the frequently used general forms (rectangular boxes, L-shaped areas, etc.). In either case it is necessary to take care that the shape of the design space should not be too complicated. Decisions about the shape of a design space can be made based on the complexity of the embedded component, the positions of the intended interface features, etc.

Interface features

The connection between the design spaces are represented by means of *interface features*. These are intended for use for two purposes:

– References for constraints. The necessary datum elements (planes, points, etc.) will be stored in the interface features.
– References for connection form features. The orientation and position of the connection features to be used to embody the physical connection will be obtained from the interface feature data.

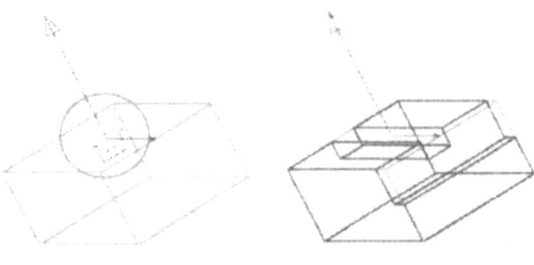

a) connection form   b) inserted component
feature attached to an   modified by the interface
interface feature: just a   feature
design space attribute

Figure 5. Concept of interface features

Since an interface feature is the reference for a constraint between two design spaces, neither designer on either sides is allowed to modify that arbitrarily. Therefore, once defined, it appears as a restriction for the designers.

Constraint classification

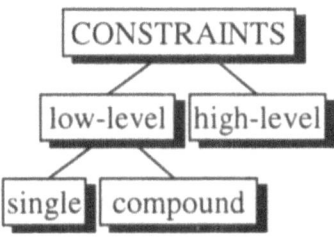

Figure 6. Constraints in the LDT

We can distinguish two fundamental kinds of constraints in the LDT: *low-level* and *high-level* constraints. In the former case the parameters are independent from each other, while the latter means there is some kind of function which connects one or more joint parameters together. The high-level constraints have, generally, only one degree of freedom.

In the case of the low-level constraints one can distinguish between single and compound constraints; the difference is whether one or more vector equations are needed to characterise the constraint. Figure 7. shows examples for the constraint types.

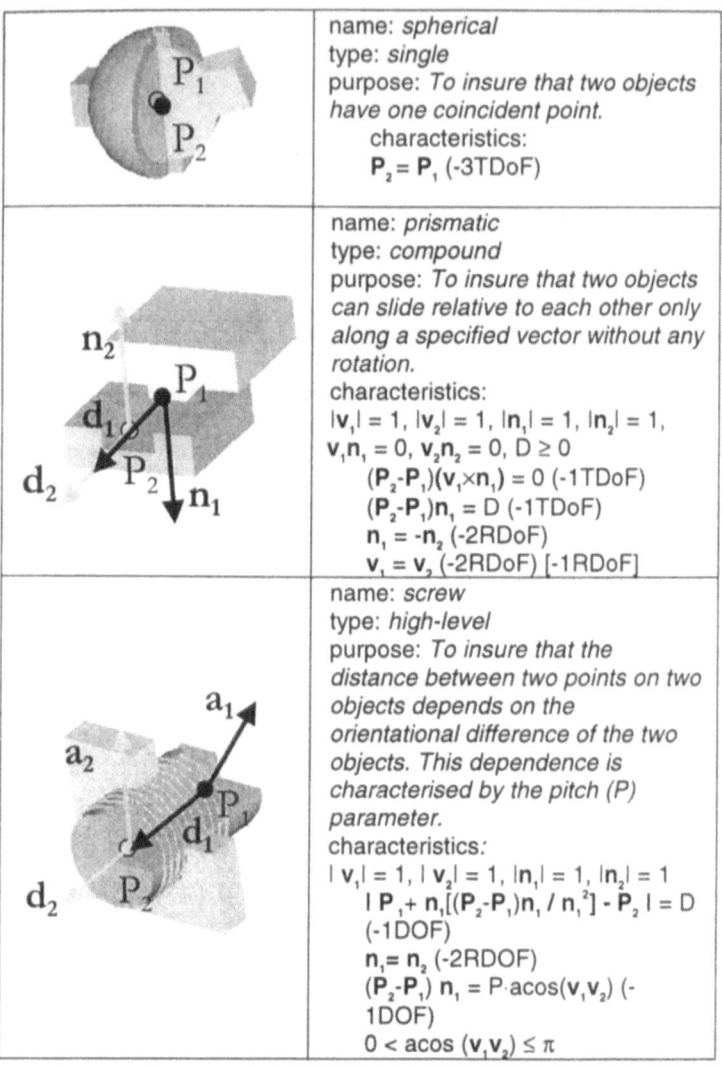

	name: *spherical*
	type: *single*
	purpose: *To insure that two objects have one coincident point.*
	   characteristics:
	   $P_2 = P_1$ (-3TDoF)

	name: *prismatic*
	type: *compound*
	purpose: *To insure that two objects can slide relative to each other only along a specified vector without any rotation.*
	characteristics:
	$\lvert v_1 \rvert = 1$, $\lvert v_2 \rvert = 1$, $\lvert n_1 \rvert = 1$, $\lvert n_2 \rvert = 1$,
	$v_1 n_1 = 0$, $v_2 n_2 = 0$, $D \geq 0$
	   $(P_2 - P_1)(v_1 \times n_1) = 0$ (-1TDoF)
	   $(P_2 - P_1)n_1 = D$ (-1TDoF)
	   $n_1 = -n_2$ (-2RDoF)
	   $v_1 = v_2$ (-2RDoF) [-1RDoF]

	name: *screw*
	type: *high-level*
	purpose: *To insure that the distance between two points on two objects depends on the orientational difference of the two objects. This dependence is characterised by the pitch (P) parameter.*
	characteristics:
	$\lvert v_1 \rvert = 1$, $\lvert v_2 \rvert = 1$, $\lvert n_1 \rvert = 1$, $\lvert n_2 \rvert = 1$
	   $\lvert P_1 + n_1[((P_2 - P_1)n_1 / n_1^2] - P_2 \rvert = D$ (-1DOF)
	   $n_1 = n_2$ (-2RDOF)
	   $(P_2 - P_1) n_1 = P \cdot acos(v_1 v_2)$ (-1DOF)
	   $0 < acos(v_1 v_2) \leq \pi$

Figure 7. Examples for LDT constraints (DoF: Degrees of Freedom, TDoF: Translational DoF, RDoF: Rotational DoF)

Creating the layout

Definition of design spaces

There are two possibilities for determining the shape of a design space. It can be either selected from a library of frequently used forms, or created by the designer. In the latter case the definition can be accomplished by determining the initial spatial configuration (position and orientation) of its reference frame, determining its faces and specifying the constraints among its geometrical elements. The newly created design space shape can be stored in a library.

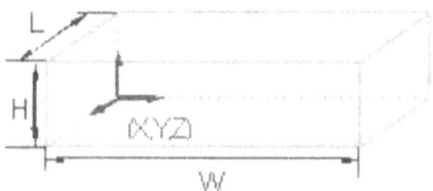

Figure 8 Parameters of a design space

Assigning interface features to design spaces

The interface features can be defined by selecting existing geometric entities, that is, lines, points, etc. It is possible to pick any entities in the model space, but all of them have to be assigned to the same design space. In other words, all datums of an interface feature will be given relative to the local co-ordinate system of the design space.

Setting up constraints between interface features

Since the interface features are tied rigidly to the design spaces, their relations to each other determine the spatial configuration and the kinematic behaviour of the layout. The layout designer has to select a joint type, point out the two interrelated interface features. The need for the user interaction comes from the fact that several constraints require the same set of geometric entities. For example, two perpendicular vectors and a point can be used as a reference for either a prismatic or a screw joint.

The relationships among the design spaces are represented by the *layout constraint graph*. The nodes of this graph are the design spaces while the edges are the joints between them. The interface object reference frames are also indicated in this graph, they appear as the connection points of the edges to the nodes. Figure 9. shows the layout constraint graph of a spatial slider-crank.

By defining the design spaces the user forms the nodes of the layout constraint graph. By setting up the interface object reference frames the connection points of the edges are created. By adding the joint type the edges are set up.

There must be a *grounded design space* which is considered to be fixed in space. The position and the orientation of this design space does not change during the kinematic analysis of the layout, but the grounded property can be added to any of the design spaces. There can be only one grounded design space in the layout, but at least one is necessary.

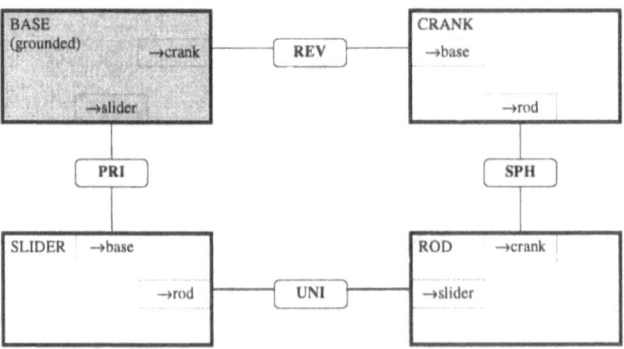

Figure 9. Layout constraint graph of a spatial slider-crank

Constraint solving

The constraint solving system uses symbolic geometric reasoning based on the work of Kramer [Kram92]. That is, the loci of interface features are calculated based on the degrees of freedom of the applied constraints and then locus intersection analysis is performed in order to find the spatial interface feature configurations where all the applied constraints are satisfied. If more than one dimensional locus intersection is found then driving inputs need to be introduced by the user in order to degenerate the intersection intervals to points or else the method stops without returning a solution. If more than one possible configuration is found the user is inquired to select the practically correct solution.

The algorithm starts processing from the fixed design space of the constraint loop to be solved. Such a design space can be either a grounded or a fully constrained design space of whose spatial configuration has already been fixed by another solver sub-process. In the next stage the process takes one of the edges of the constraint loop from the fixed design space and calculates the loci subsequently along the selected path. If a surface locus is found the algorithm stops processing the current branch and continues with the other one. In the optimal case the intersection of the tail loci of the two branches results in discrete points so that the user can select a practically possible solution. Otherwise the user is asked to introduce driving inputs.

Two examples are shown here in order to illustrate the operation of the constraint solving algorithm. In the first case a spatial slider-crank mechanism is analysed. Its model can be seen in Figure 10.

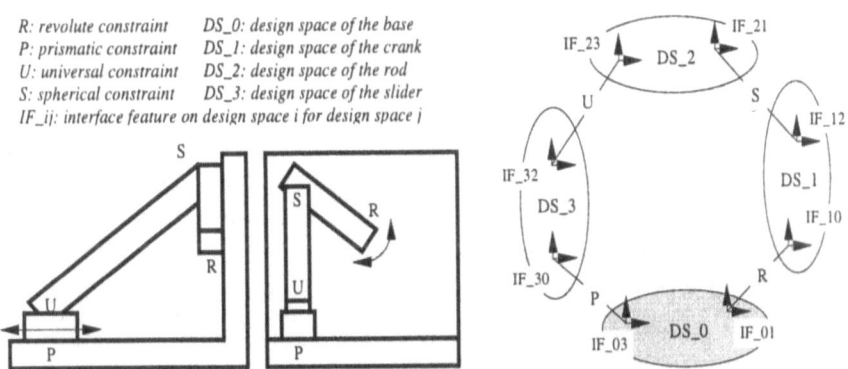

Figure 10. Spatial slider-crank example

The solver starts from DS_0 and calculates the locus of IF_10 which is a point. (In the current implementation of the algorithm the solver selects the branch which continues in the lower index design space.) At IF_12 a circle locus is found such as at IF_21. In the next step the solver leaves this branch because the locus of IF_23 is a torus. Since this kind of locus will surely result in a non-discrete set of solutions at the end, it makes no sense to continue with the other branch, instead the user is asked to introduce a driving input. In this practical case this means to set the parameter of the revolute constraint to a certain degree. With this new configuration the locus of IF_23 is a spherical surface, so the process can continue with the other branch. Here the prismatic and universal constraints result in a linear locus for IF_32. The intersection of this line and the recently calculated spherical surface creates n intersection points where $n \in \{0,1,2\}$. If $n=2$ then the user has to choose a solution. After the intersection point is fixed the two parameters of the universal constraint will be set by the system. If $n=0$ the geometry of the layout is to be modified.

The second example is an "ordinary" vice assembly (see Figure 11).

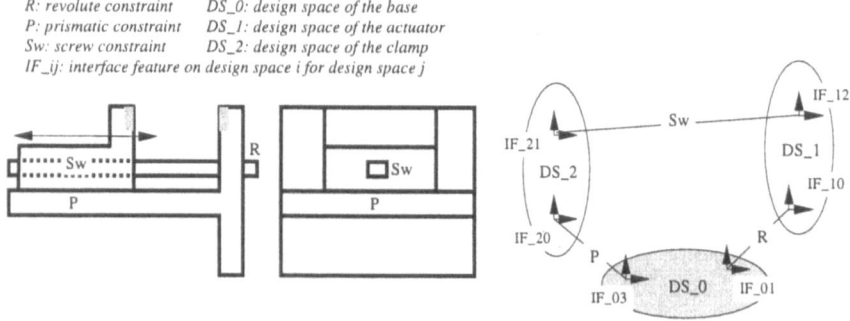

Figure 11. Vice example

The first relevant locus (starting with the revolute constraint) is a helical curve at IF_20. From the other side IF_20 must be on a line. The intersection of these loci can result in a discrete set from which the user has to select a practically acceptable solution.

Kinematic analysis of the layout

Once the design spaces have been created and the necessary interface features are adjusted, the layout can be arranged in space by invoking the analytical solver. At the end of this phase the system investigates if the resulting layout structure is rigid or not. If not, the non-rigid constraints will be pointed out and the user will have the possibility of assigning driving inputs to the relevant parameters. By altering a driving input with small values or defining functions, the layout designer can trace the kinematic behaviour of the layout (Figure 12.).

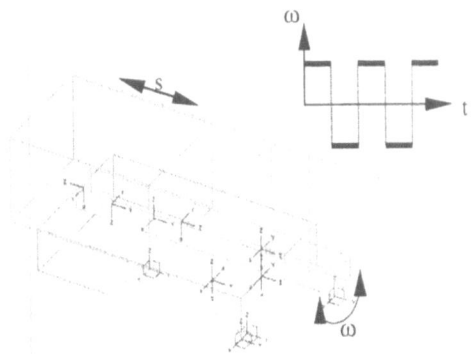

Figure 12. Early stage kinematic analysis of a vice.

The refinement stage

There are various possibilities open for designing the part. Having set up the connection structures the designer has, predefined, a geometrical framework within which to work. An important part of this is the connection form feature set defined by the layout designer. The designer might choose to start with these features and build the design with reference to these or also edit them into the design at or near the end of the design process. Note, though, that the connection features represent a concrete realisation of the abstract connection defined by the layout designer. In reality they are pairs of positive and negative elements. Keeping them as higher level abstract form means that the designer can negotiate, over which form, the positive or negative, he includes in his design.

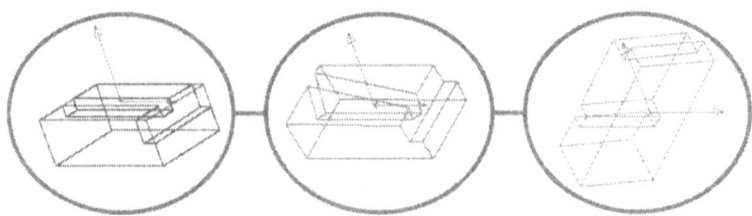

Figure 13. The orientation and the position of a connection form feature depends only on its reference interface feature regardless of the component to which it is to be attached

A design space is the responsibility area of its unit designer. From the point of view of the layout the shape and the complexity of the component is irrelevant, the only restrictions are the attached interface features. The unit designer must use them as references for his connection form features in order to keep the assembly realisable.

An example

In the following section an example for the layout design process will be presented in order to illustrate the operation of the LDT. This example (Figure 14.) has been realised by the first working demo of the LDT.

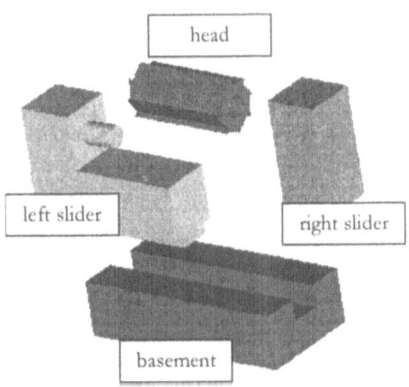

Figure 14. Milling head assembly example

Introducing design spaces

Each of the design spaces contains only one part (Figure 15.). The basement DS, the right slider DS and the head DS are of the "cuboid" type, while an "extruded" design space has been defined for the left slider. In the latter case one has to define a set of coplanar points and then extrude the resulting section by adding a thickness value to it.

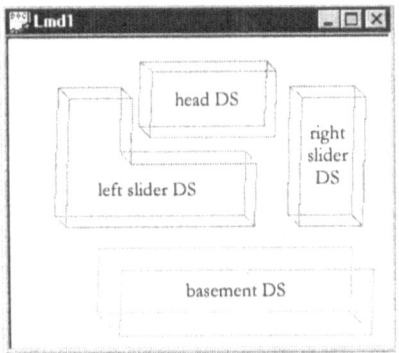

Figure 15. Design spaces

Introducing interface features

The interface features introduced are shown in Figure 16. These are in fact vectors and points in this case, but these can be easily hidden by means of a graphical user interface in order not to force the user to deal with low-level geometric entities. There are two interface features for the <basement ⇔ left slider> connection, two for the <right slider ⇔ left slider> connection and two for the <left slider ⇔ head> connection. (see Figure 17.)

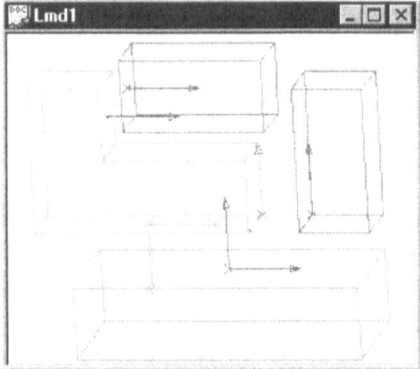

Figure 16. Interface features (datums)

Introducing constraints

Three constraints are introduced: the left and the right slider are attached rigidly to each other ("rig"), the left slider and the basement use a prismatic slider connection ("pri"), while there is a revolute joint ("rev") established between the left slider and the head (Figure 17.).

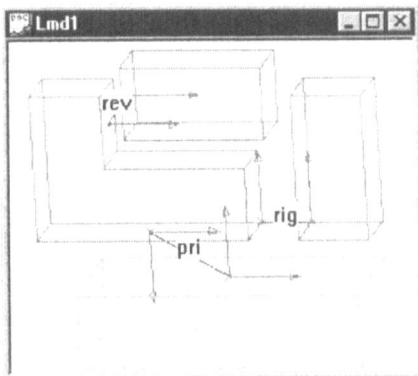

Figure 17. Constraints

Kinematic analysis I.

At this point the analytic constraint solver can be invoked to put the layout together. In order to do this the user has to specify a design space as grounded. The system can detect if there are remaining degrees of freedom and a driving input can be set up to one of these free parameters. By modifying the driving input value(s) the kinematic behaviour of the layout can be investigated. Moreover, collision checks of the design spaces can be performed, but this operation requires further research.

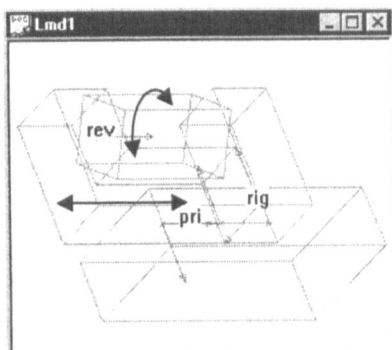

Figure 18. The assembled layout and its possible movements

Introducing components

Up to now only the layout design is done which is the responsibility of the unit designer. Now the geometric refinement work is distributed among several unit designers who can work parallel to each other, as opposited to the sequential manner of the layout design process.

Four components are inserted into the design spaces: the head part, the left slider part, the right slider part and the basement part (Figure 19.).

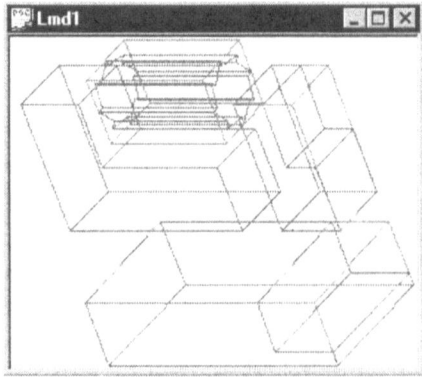

Figure 19. Components in the design spaces

Introducing connection form features

The interface features determined by the layout designer act as restrictions for the unit designers. That is, they have to determine the connection form features of the components by keeping the interface features in view.

In this case the sliding joint is realised by a slot (basement) and a rib (left slider), the rotational relationship is realised by a pin (left slider) and a through hole (head) form feature. For simplicity we do not assign any connection form feature to the rigid connection, as though the two parts were welded together. The resulting components can be seen in Figure 20.

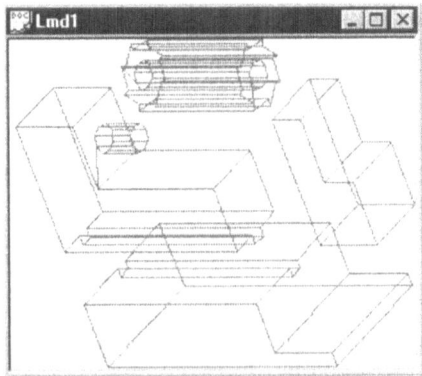

Figure 20. Components modified by connection form features

Kinematic analysis II.

Similar to the kinematic analysis at the end of the layout design stage (section 0) new movement and collision checks can be performed with respect to the components (Figure 21.).

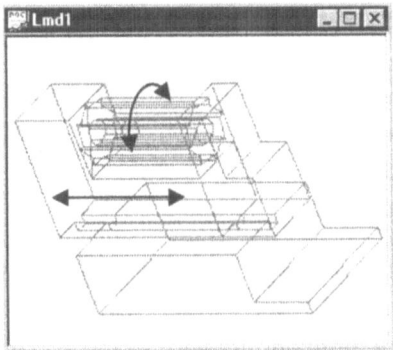

Figure 21. The result of the design process

Open issues

Several problems may emerge from the approach presented, some of which are summarised in this section.

Static intersection detection.

How should a unit designer consider a design space which intrudes into his? The need for this analysis comes from the fact that a spatial area can be occupied by only one component at the same time, so one has to decide on the distribution of the common areas. In most cases the decision can be made based on the connection form features to be applied or the position and the orientation of the interface objects.

Dynamic intersection detection.

What happens if two design spaces collide with each other during the kinematic analysis? The layout designer has to examine the possible intersections and decide on which of the relevant unit designers is forced to provide space for the intersecting design space and in which degree. This task can be neglected if the final assembly to be designed is a rigid structure.

Subdivision of design spaces.

If the unit designer is about to design a subassembly he can involve more designers in the work by dividing his design space up into lower level design spaces. This way he becomes the layout designer of the subspaces, but from the other side he may have pure components in his design space as well. The question of this mixing of design spaces and components at the same level is under consideration.

Shape modifications of design spaces and components.

In most cases there is a need for changing the shape and the dimensions of a design space. For example, the layout designer detects a non-allowable collision during the movement check or one or more dimensions need to be changed automatically in order to keep the layout consistent (Figure 22, Figure 23). There is another question, namely what would happen with the component inside, how can the parameters of the components follow the changes of their design spaces.

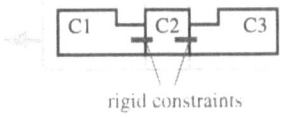

rigid constraints

Figure 22. User action: translation of design space DS1 (DS: Design Space, C: Component)

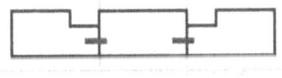

Figure 23. The L dimension of DS2 (C2) is adjusted automatically.

Access rights

There are several ways to assign layout entities to the groups and individuals, but in most cases the entities cannot be assigned to the designers in a strictly separated manner. For example, there might be a need to modify the dimensions of a design space by the unit designer for design optimisation purposes, but this may need to be done by the layout designer as well because of intersection problems during kinematic analysis. As another example, a screw fastener cannot be assigned conceptually to any of the designers of the components to be put together with the screw, they should have equal rights to manipulate the screw, if necessary.

Several problems might emerge from the shared property of the elements of the layout. Therefore, some kind of mechanism is needed to keep track of, and control the operations to be performed on them. In the LDT access records connected to the layout entity objects are to be used to describe who can access and influence the entities and how. The access records should not be attached rigidly to the layout, because the access management routines are developed independently from the layout entity classes. In this way the whole access control system can be replaced by detaching the access records; even by using on-site developed software components. This work is still in the preparation stage.

User interface facilities.

In order to facilitate the fast modification of the layout several visual processing issues are to be solved. For example the direct manipulation of control points applied by van Emmerik [Emme91] could also be applied here, but there are several cases when there is a need for pointing out edges, facets, etc. During the kinematic analysis it is a good idea to tie the mouse movement and a free parameter together.

Since the orientation and the position of the connection form features are predetermined by the interface features, only the feature parameters are needed to be set up during the geometric refinement. In addition, standardised features are often used, therefore the feature operations could be done in a "cut-and-paste" manner (by using the principles discussed in [RaIn93]).

Conclusion

The approach presented can be used to design the layout of complex assemblies from scratch without having any detailed geometry at the level of the physical components. The system in its current state is able to introduce design spaces, interface features and constraints, thus forming the first draft of the assembly to be realised. An analytical constraint solver is used to put together the layout elements, the kinematic analysis of the assembled layout can be done with the same solver. After this first preparation stage of the layout concurrent work on different components can be started by distributing refinement work among the unit designers. Since the relationships are established and maintained by the layout designer the geometry of the components does not influence the kinematic behaviour of the assembly, therefore, the modification of the layout can be done in parallel to the detailed design.

Numerous questions and problems arose from the computer implementation of this method, some of them have been mentioned in section 0. In addition, an easy-to-use graphical user interface has to be created in order to hide the low-level operations from the user. The analytical solver used must be improved or replaced in order to achieve better performance. Finally, the whole system has to be integrated in an existing design environment and the facilities of networked concurrent work should be exploited.

Acknowledgement

The authors are indebted to Dr. Ian Stroud, who contributed so much to the development of the LDT prototype and the preparation of this paper.

References

[KuDe96]   Kuttner, B., Deitz, D.: Reviewing designs in cyberspace; Mechanical Engineering, dec. 1996, p56-58

[GuSt96]   Guan, X., Stevenson, D. A., MacCallum, K. J.: A prototype system for early geometric configuration design Computers in Industry 30, 1996, p233-239

[LaFe96]   Lashin, G., Feldhusen, J.: A CAD-based tool for development of large layouts Research in Engineering Design 8, 1996, p217-228

[SeMe96]   Seybold, B., Metzger, F., Ogan, G., Bathelt, J., Collenberg, F. Taiber, J.,Simon, K., Engeli, M.: Spatial modelling with geometric constraints Technical Report for Swiss industrial project KTI 2726.1., 1996

[ScSc96]   Schmidt, R., Schmidt, M. (eds.): Computer Aided Concurrent Integral Design, Project 5168, ISBN 3-540-60480-4

[LeJo96]   Leigh, J., Johnson, A. E., Vasilakis, C. A., DeFanti, T. A.: Multi-perspective collaborative design in persistent networked virtual environments, 1996
http://evlweb.eecs.uic.edu/spiff/calvin/calvin.vrais/index.html,

[RaIn93]   Ranta, M., Inui, M., Kimura, F., Mäntylä, M.: Cut and paste based modeling with boundary features; 2nd ACM Solid Modeling, Montreal, 1993, p303-312

[Stro93]   Stroud, I.: Modelling techniques for handling non-geometric information 2nd ACM Solid Modeling, Montreal, 1993, p367-376

[Kram92]   Kramer, G. A.: Solving geometric constraint systems; MIT Press, 1992, ISBN 0-262-11164-0

[Emme91] van Emmerik, M. J. G. M.: Interactive design of parametrized 3D models by direct manipulation PhD. thesis, Delft University Press, 1991

[Haug89] Haug, E. J.: Computer Aided Kinematics and Dynamics of Mechanical Systems, Allyn and Bacon, 1989, ISBN 0-205-11669-8

Solving Over- and Underconstrained Geometric Models

Alex Noort, Maurice Dohmen, and Willem F. Bronsvoort[*]

New approaches to solving underconstrained and inconsistent overconstrained 3D geometric models are described.

When a model is found to be underconstrained, the underconstrained parts of the model are determined and shown to the user. The user has to add constraints to these parts, to make the model well-constrained.

When an inconsistency in the model is found, the set is determined of the constraints that together cause the inconsistency. The model is made well-constrained, either by modification of one of these constraints, or by removal of one of them from the model.

The approaches make use of a dependency graph, which contains information about dependencies between model variables. When an inconsistency is detected, the set of conflicting constraints is determined by traversal of the dependency graph. When an underconstrained situation is detected, the model variables in the underconstrained parts of the model are determined using this graph.

The concept of storing information about dependencies between model variables in a graph, can be used with all non-iterative constraint satisfaction techniques. An implementation of the approaches has been made in a 3D constraint solver based on degrees of freedom analysis. It identifies rigidly connected parts of the model, and then solves the geometric constraints in these parts incrementally by geometric transformations, taking into account the translational and rotational degrees of freedom of the constrained model parts. The solver handles loops of constraints by rewriting some of the constraints in the loop.

---

[*] Faculty of Information Technology and Systems, Delft University of Technology

Introduction

In *constraint-based feature modeling, features* are the basic modeling entities. Features contain *shape* and *non-shape* information. The non-shape information can be information about the function of the feature, the manufacturing or assembly of the part the feature belongs to, etc. Both shape and non-shape information are used to assist the user when editing the model, by ensuring the validity of the features in the model.

Both types of information are represented by *constraints*. To satisfy these constraints, they are solved by a *constraint solver*. Here we concentrate on geometric constraints, such as a constraint that specifies two planes to be at a given distance of each other. These constraints represent shape information of features, such as the parameters, i.e. position, orientation and dimension.

In the feature modeling system SPIFF [BBDHK97], *high level* constraints are mapped to combinations of *low level* constraints which are solved by the constraint solver. Shapes contained in the features used in SPIFF are mapped to low-level geometric objects that the constraint solver can deal with. The constraint solver is based on *degrees of freedom analysis* [Kra92].

The models can only be solved if they are *well-constrained*, which means that there are exactly enough constraints in the model to uniquely determine the parameters of all features in it. When some parameter is determined in different ways by multiple groups of constraints, the model is called *overconstrained*. When some parameter is not determined, the model is called *underconstrained*.

In the following sections, first the concept of degrees of freedom analysis on which the constraint solver is based is presented. After that, the *dependency graph* that is used to remove under- and overconstrained situations from the constraint model is presented. Then, the ways over- and underconstrained situations can be repaired, i.e. removed from the model, are presented. Finally, results and conclusions are presented.

Although the concept of a dependency graph containing information about dependencies between model variables is presented here for a geometric constraint solver based on degress of freedom analysis, it can be used with all non-iterative constraint satisfaction techniques.

Degrees of freedom analysis

This section describes the concept of degrees of freedom analysis [Kra92], on which the geometric constraint solver has been based. Degrees of freedom analysis is a structured constraint solving approach. It solves a constraint model bottom-up, by first solving and then combining subgraphs using geometric knowledge. Constraint models consist of *geoms* with *markers* that represent the geometric elements of the model, and geometric *constraints* between markers of different geoms.

Geoms contain a coordinate system that has a variable orientation and position with respect to a world coordinate system. Markers are represented by a position and an orientation, and are fixed with respect to the geom they are part of. The set of all constraints directly between two geoms is called a *joint*. A joint is called *rigid* if the position and orientation of one geom of the joint are completely determined relative to the other geom of the joint by the constraints in the joint.

Every geom initially has three *translational degrees of freedom* and three *rotational degrees of freedom*, from now on called TDOFs and RDOFs. These degrees of freedom describe the directions in which a geom can move and rotate without violating any constraint already satisfied. The (R/T)DOFs of a geom are mutually independent, i.e. perpendicular.

A *constraint graph*, consisting of geoms, markers and constraints, is used to represent a model to be solved. This constraint graph contains geoms as nodes and joints as edges. A constraint graph is solved by merging geoms that have a rigid joint between them into a *macro-geom*; the graph has been solved when there is only one geom left.

The constraint solver alternates between two tasks. The first task is to solve constraints of rigid joints, and to merge the geoms of these joints into macro-geoms. The constraints are satisfied using action analysis. The second task is to rewrite constraints from one joint to equivalent ones in other joints of the model, in order to introduce new rigid joints. This is done using locus analysis.

Action analysis solves the constraints in a joint incrementally. One geom of the joint is taken fixed, and the other one is allowed to move. The constraints of the joint are solved one by one, by determining the

DOFs left of the free geom, and then reducing these DOFs according to the constraint to be solved. The DOFs are reduced in such a way that the free geom will obey the constraints already solved.

The new DOFs of the free geom and the transformation needed to satisfy the constraints are registered in so-called *plan fragments*. For every DOF configuration of a free geom and a constraint type to be solved, such a plan fragment has been implemented.

Locus analysis is performed between three geoms; in this process, constraints from a joint between two of these geoms are rewritten to the joints on the third geom, see Fig. 1. Before locus analysis can be performed, action analysis is done on the joints of the third geom, i.e. the joints with the marked constraints in Fig. 1(b), with this geom being fixed. After action analysis has been performed, the two geoms of the joint on which locus analysis will be performed have certain DOFs left with respect to the third geom. Locus analysis will then be performed on the joint between these geoms, i.e. the joint with the marked constraint in Fig. 1(c).

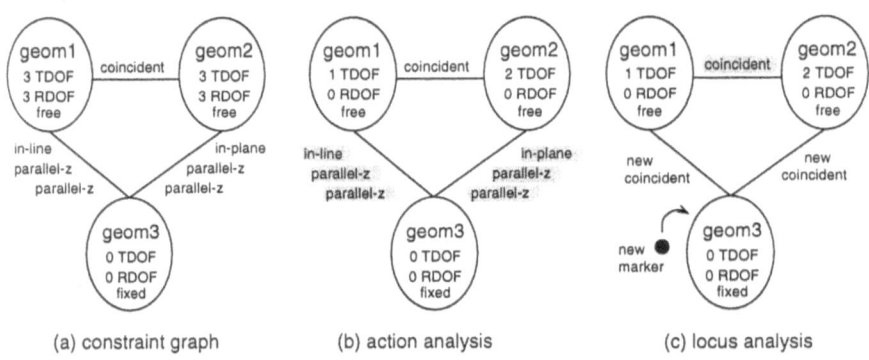

Fig. 1: Locus analysis on three geoms.

A locus is the area that can be reached by a geometric primitive, when moving according to its degrees of freedom. In locus analysis, the intersection of the loci of the markers of a constraint is the area where the markers have to be to satisfy the constraint. According to the intersection of the loci, new markers and new constraints are added to the model. These new constraints reduce the same DOFs between the two geoms of the original constraint as the original constraint does, but via the third geom.

In Fig. 1(c), an example of locus analysis is given for a coincident constraint between a geom with one TDOF left and a geom with two TDOFs left after action analysis. The DOFs of these geoms are perpendicular to each other. The locus of the marker on the geom with one TDOF, is a line parallel to the DOF line and through the marker. The locus of the marker on the geom with two TDOFs, is a plane parallel to the two DOF lines and through the marker.

The intersection of those two loci is the point where both markers have to be in order to satisfy the coincident constraint. A new marker is added to the third geom, with as position the intersection of the loci, and two coincident constraints are added between this new marker and the markers of the original constraint. These two constraints will reduce the same DOFs between the two geoms of the joint on which locus analysis has been performed as the original constraint.

As shown in the example of Fig. 1, locus analysis can be used to solve loops with three geoms. To solve a loop with more than three geoms, it has to be reduced to a loop with three geoms. This is done by repeatedly replacing pairs of consecutive joints by a joint between the first and the last geom of the pair, thus removing a geom from the loop until there are three geoms left, see Fig. 2.

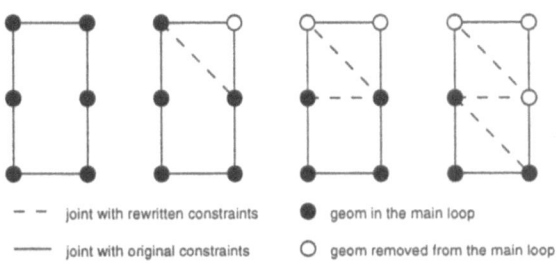

- - joint with rewritten constraints   ● geom in the main loop

── joint with original constraints   ○ geom removed from the main loop

Fig. 2: Splitting a loop with more than three geoms.

The approaches to deal with over- and underconstrained models that have been presented before by others are described in the next paragraphs. In dealing with over- and underconstrained models, three phases can be distinguished. In the first phase, detection takes place. For an overconstrained model, the *involved constraints*, i.e. the constraints that cause the inconsistency, are detected. For an underconstrained model, the undetermined properties are determined.

In the second phase, a constraint needs to be chosen to repair the over-
or underconstrained situation. For an overconstrained model, a
constraint needs to be chosen to be removed. For an underconstrained
model, a constraint needs to be chosen to be added. In the third phase,
the model needs to be re-solved.

To be able to detect an over- or underconstrained situation, the
constraint solver has to provide facilities to register the geometric
properties determined by individual constraints. In this way, the solver
is able to find the constraints involved in an overconstrained situation.
Furthermore, the solver must be able to partly solve a model in order to
find the geometric properties not determined in the model. Such
facilities can only be provided by structured, non-iterative constraint
solvers such as used in [SAK90,KKP95,HB97].

To repair an overconstrained situation, a constraint has to be
removed or edited. Removing a constraint really removes the
overconstrained situation, whereas editing a constraint turns the
inconsistent overconstrained situation into a consistent one. The choice
of a constraint to be removed or edited can be made using several
criteria. An example of such a criterium that is used by many systems,
is the user's knowledge of its design goal, i.e. let the user choose a
constraint [SAK90]. A simple extension of this is to prohibit the user
from choosing constraints of a particular type, e.g. constraints that
determine topological relationships [KKP95]. Removing topological
relationships is often not desirable, because it violates the integrity of
the shapes of the model.

To repair an underconstrained situation, a constraint to be added can
be chosen using default constraints [SAK90]. In this approach, for each
type of DOF, left between two geometric elements, a constraint of a
default type is inserted into the model to reduce the DOF.

Dependency graphs

This section describes the dependency graph, a data structure used to
repair over- and underconstrained situations in a model. When the
constraint solver has found an overconstrained situation, it should be
able to find the constraints involved in this situation. These are the

constraints that together reduce the same DOFs between two geoms as the constraint for which the inconsistency was detected. To find these constraints, the constraint solver has to know the DOFs reduced by each constraint. This *dependency information* is known when the constraints are solved, and can then be stored.

Dependency information is stored in dependency graphs that consist of geom nodes that represent the geoms of the model and constraint nodes that represent the constraints that reduced DOFs when the model was solved. In Fig. 3, an example of a model and its dependency graph is given. In this example, the orientation and translation along the z-direction of the blocks have already been fixed. The constraint nodes contain the DOFs that the constraint reduced in solving the model, i.e. the *reduced DOFs*, and the *orientation* of the constraint, i.e. the orientation of the relevant marker of the constraint. The DOFs are denoted by T(x,y,z) for translation along a line with (x,y,z)T direction, and R(x,y,z) for rotation about an axis with $(x,y,z)^T$ direction; the orientation of the constraint is denoted by O(x,y,z). This orientation is used to be able to determine the *initially reduced DOFs* or *initial DOFs*, i.e. the DOFs a constraint reduces when no other constraint is involved. These initial DOFs depend on the constraint type and the orientation. The reduced DOFs differ from the initially reduced DOFs if the initial DOFs of the constraints of the joint are dependent.

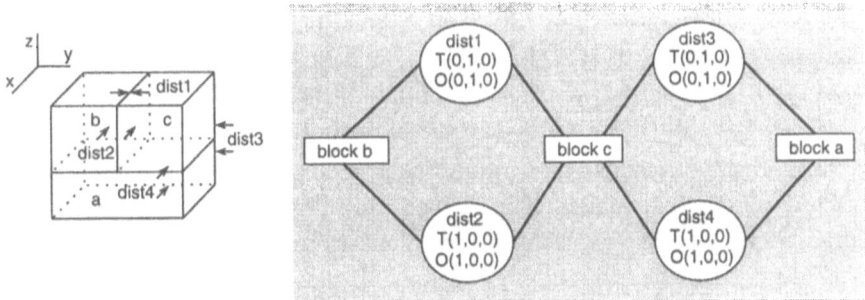

Fig. 3: An example of a model and its dependency graph.

In Fig. 4, an example of the initial DOFs of a constraint and its reduced DOFs in a joint with multiple constraints is given for a model in which the initial DOFs are dependent. The initial DOF of the dist1 constraint is T(0,1,-1), as can be found in the dependency graph of the first model.

The initial DOF of the dist2 constraint of the second model is T(0,0,1). Because the initial DOFs are dependent, and the dist2 constraint happened to be solved first, the reduced DOF of the dist1 constraint differs from its initial DOF.

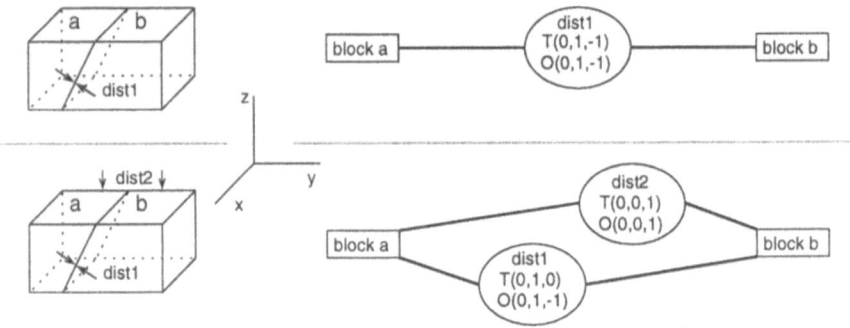

Fig. 4: A model that shows the initial DOFs and one with dependent initial DOFs.

A model in which constraints exist whose reduced DOFs contain DOFs not initially reduced by the constraint, is called *non-aligned*. An *aligned* model is a model of which the constraints reduce only DOFs that they reduce initially. This information is used when the dependencies between two geoms are determined, because in a non-aligned model, geoms do not only depend for a given DOF reduction on the constraint reducing this DOF, but also on the constraints that enable the constraint to reduce the DOF. The model of Fig. 3 is aligned, but the second model of Fig. 4 is non-aligned, because the dist1 constraint reduces T(0,1,0) that it does not reduce initially.

The fact that constraint nodes represent constraints that reduced DOFs when the model was solved, implies that some constraints involving locally consistent overconstrained situations are not registered in the dependency graph. In such a situation, multiple constraints can reduce the same DOF. When solving, one of these constraints is used to reduce the DOF and is registered in the dependency graph, whereas the other constraints are not. These other constraints will therefore not be found when searching involved constraints.

An example of this situation is given in Fig. 5, where the dist2 and dist3 constraints form a consistent overconstrained loop. Because of this, one of these constraints, dist3, will not reduce any DOFs, and will therefore not be registered in the dependency graph. In case an inconsistent overconstrained situation will be found later that involves this consistent overconstrained situation, this constraint will not be recognized as an involved constraint. Thus the overconstrained situation will not always be removed when removing an involved constraint from the model.

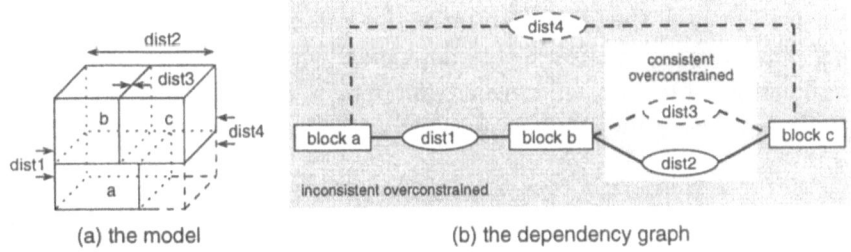

(a) the model   (b) the dependency graph

Fig. 5: Locally consistent overconstrained and globally inconsistent overconstrained.

Our dependency graph is similar to the dependency graph as used in the system described by Keirouz et al. [KKP95], whose constraint solver is also similar to the one used in this system. The main difference between the approaches is that geometric elements in their constraint graph are fully determined by one constraint, whereas geometric elements in our constraint graph need multiple constraints.

Another application of dependency graphs is described by Hsu and Brüderlin [HB97]. The similarity between their and our application is that both involve the DOFs reduced between geometric elements of the model. The difference is that in the application described by Hsu and Brüderlin, the dependency graph is used to determine the part of the model to be re-solved after a geometric element has been changed, whereas in the application described here, it is, among other things, used to find constraints involved in an overconstrained situation. Therefore, the dependency graph used here does not only register the number of DOFs reduced between pairs of geoms, but also which DOFs are reduced between them.

Dependency graphs are built in two phases for each macro-geom that is created during the solving of a model. In the first phase a *joint graph* is built, and in the second phase this joint graph is expanded into an *expanded graph*. These two phases correspond to the way the solver works. The solver first solves the constraints between two geoms, and after that, in case the geoms have become rigidly connected, the solver merges them.

In the first phase, the solver does not change the structure of the constraint graph for the geoms of the joint. It only reduces the DOFs of the free geom of the joint. Similarly, the building of the joint graph does not change the structure of the dependency graphs of the geoms of the joint either, but only adds constraint nodes to the joint graph of this joint, containing the DOFs reduced by these constraints, see Fig. 6(a).

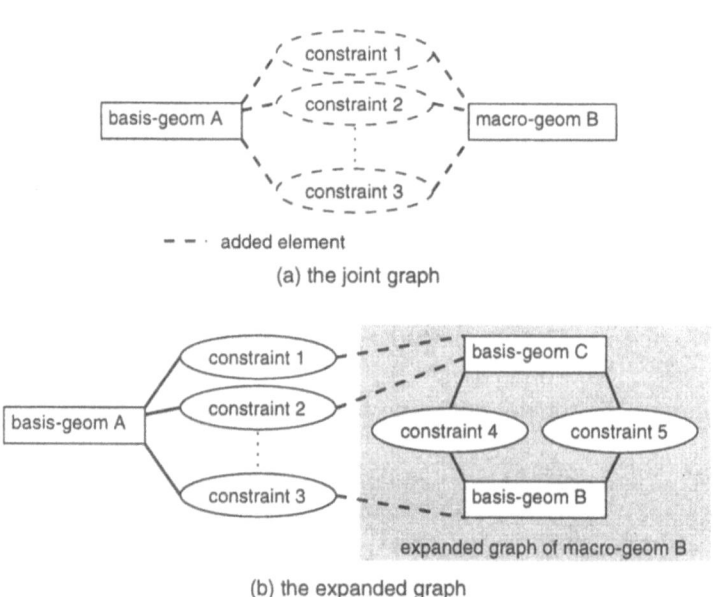

Fig. 6: Building dependency graph in two phases.

In the second phase, the solver does change the structure of the constraint graph by merging the two geoms into a new macro-geom. The expanded graph built during this phase is created by replacing the macro-geom nodes of the joint graph by the expanded graphs representing these macro-geoms, see Fig. 6(b). Every macro-geom in the model has its own expanded dependency graph, describing the dependencies between all *sub-geoms* of that macro-geom.

These sub-geoms are all *basis-geoms*, i.e. geoms that are present in the constraint graph when the solving is started. In case an overconstrained situation is found in a joint on a macro-geom, the expanded dependency graph can be used to find all constraints that reduce given DOFs between given sub-geoms. In case an underconstrained situation is found, the expanded dependency graphs of the macro-geoms are used to determine the basis-geoms that are fixed with respect to each other.

Overconstrained models

This section describes the detection and repairing of an overconstrained situation in a model. An overconstrained situation is detected by the constraint solver when it cannot determine a geometric transformation for a geom given its DOFs in order to satisfy the constraint that is solved. In such a case, another constraint has already removed DOFs that the constraint solver needs to satisfy this constraint. These DOFs are called the *overconstrained* DOFs. The set of involved constraints is built from the constraint that could not be solved, and the constraints that reduced the overconstrained DOFs between the two geoms. The overconstrained situation is repaired by choosing one of the involved constraints to be adjusted or removed from the model.

The constraints involved in an overconstrained situation are found in the dependency graph. To find them, the overconstrained DOFs need to be known first. These DOFs are detected by action analysis when it moves the free geom in order to satisfy the constraint, and by locus analysis when it tries to intersect the loci of the markers. After these DOFs have been detected, the other constraints that reduce these DOFs between the two involved geoms have to be found. These constraints are found using the expanded dependency graphs of the macro-geoms and the joint graphs of the treated joints. When searching in the dependency graphs, two cases have to be distinguished: searching in an aligned model and searching in a non-aligned model.

When the model is aligned, all involved constraints are found by searching for the constraints that reduce DOFs not perpendicular to the overconstrained DOFs between the two geoms. An example of this is given in Fig. 7. This model consists of three fixed size blocks. The relative orientation and position along the z-axis of these blocks are fixed from the beginning, and the other DOFs are reduced by the distance constraints.

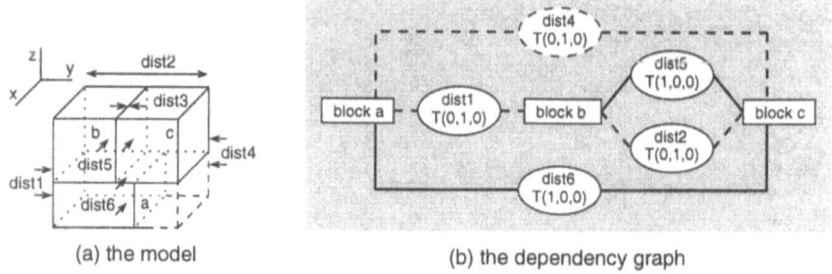

(a) the model   (b) the dependency graph

Fig. 7: Aligned model with its dependency graph. All blocks in the model have fixed dimensions.

Let us assume that the constraint that could not be solved in this model is the dist4 constraint. The overconstrained DOF of this overconstrained situation is the translation DOF in the $(0,1,0)^T$ direction. The involved constraints are found in the dependency graph by finding the paths between the two geoms block a and block c that consist of constraints reducing DOFs not perpendicular to the involved DOF $(0,1,0)^T$. In this graph there is only one such path, which consists of the dist1 and the dist2 constraints, see Fig. 7.

When the model is non-aligned, the technique used to find the involved constraints in an aligned model does not return the complete set of involved constraints. In a non-aligned model, besides the constraints that reduce DOFs not perpendicular to the overconstrained DOFs, other constraints can also be involved in the overconstrained situation. These other constraints are found by an extension of the approach for an aligned model.

For each of the constraints found by the algorithm for an aligned model, the initial DOFs are determined, see the section on dependency graphs. When the reduced DOFs of such a constraint contain DOFs that are not initially reduced by the constraint, additional constraints are involved in the overconstrained situation.

To find these additional involved constraints, first the reduced DOFs causing this extra reduction need to be determined. These are the reduced DOFs that together with the DOFs reduced by the constraint can reduce the initially reduced DOFs of the constraint. After that, the constraints reducing these DOFs between the two geoms are searched for, and added to the involved constraints.

For an example of the repairing of an overconstrained situation for a non-aligned model, see Fig. 8. In this model, the relative orientation and translation in the x-direction of the blocks have already been fixed. The model is similar to the model of Fig. 7(a), except that blocks b and c now have two parallel oblique faces. Because of this, the dist4 constraint reduces a DOF that it does not reduce initially, which makes the model non-aligned.

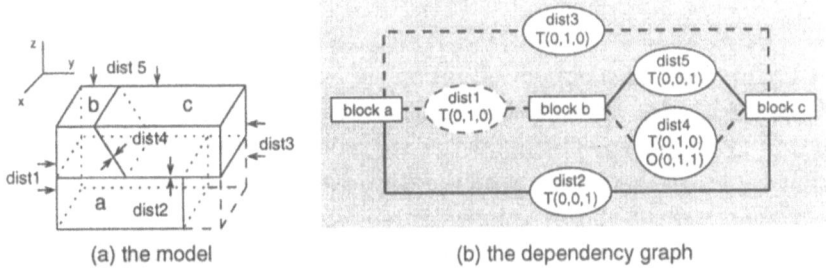

(a) the model   (b) the dependency graph

Fig. 8: Non-aligned model with its dependency graph. The dimensions of the blocks are fixed.

Let us assume that an overconstrained situation has been found when constraint dist3 could not be solved. The overconstrained DOF is the translation in the $(0,1,0)^T$ direction. The involved constraints are found by first finding the constraints in the path between block a and block b that reduce this direction. This path is denoted by dashed lines, see Fig. 8(b).

One of these constraints, the dist4 constraint, reduces a DOF that it does not reduce initially. Therefore additional constraints have to be found that enable this constraint to reduce the DOF. These constraints need to reduce TDOFs in a direction such that the combination of these

TDOFs and the reduced DOFs reduces the initial DOFs of the dist4 constraint. In this example, a translation reduction dependent on the $(0,0,1)^T$ is needed, see Fig. 9. Therefore, the dist5 constraint is also an involved constraint in this overconstrained situation.

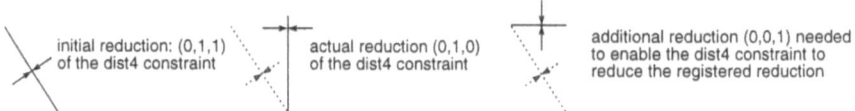

Fig. 9: Additional reduction needed to reduce registered reduction.

When the involved constraints have been found, one of them has to be chosen to be replaced or edited. Replacing a constraint involves constructing a constraint that only reduces the DOFs that are reduced by the chosen constraint that are not overconstrained, and replacing the chosen constraint by this new constraint. Editing a constraint involves adjusting the constraint parameters in such a way that the constraint is no longer inconsistent with the other constraints in the model.

If an overconstrained situation is to be repaired by editing a constraint, then the way the constraint parameter has to be adjusted needs to be known. This adjustment can be derived from the constraint that could not be solved, which is the dist3 constraint in Fig. 8(b). The parameter of the chosen constraint, e.g. the dist1 constraint, needs to be edited in such a way that the dist3 constraint is satisfied after the parameter has been changed.

The distadvantage of editing a constraint is that the model is still overconstrained. When later the parameter of a constraint involved in this overconstrained situation is changed, an inconsistency can again be introduced. When the constraint is replaced, the overconstrained situation is removed from the model, so no inconsistency can be introduced by changing a parameter of a constraint.

When the chosen constraint has been edited or removed, part of the model has to be re-solved to propagate the changes. Currently, the part of the model to be re-solved is the part of the model that is represented by the expanded dependency graph that contains the chosen constraint. A better strategy would be to re-solve only the smallest part of the model that depends, for its position and orientation relative to other parts of the model, on the chosen constraint. This part can be determined using the dependency graph that contains the chosen constraint.

Underconstrained models

This section describes the detection and repairing of an underconstrained situation from a model. An underconstrained situation has been found by the constraint solver when not all geoms can be merged, because of one or more non-rigid joints. In this situation, multiple groups of geoms exist in the model that do not have a fixed relative position and orientation. The relative DOFs between two of these groups can be found by fixing one of them, and then solving all constraints between them using action analysis. The geoms of these groups are found by traversing the dependency graphs of the macro-geoms representing these groups.

Action analysis, however, solves constraints per joint, and is therefore not able to determine the relative degrees of freedom of two geom groups when multiple paths between them exist in the constraint graph. In such a case, a new joint, directly between the two groups, is added to the model. This joint introduces a loop in the model, and the locus analysis algorithm, see the section on degrees of freedom analysis, is used to rewrite constraints from the other joints of the loop to the new joint.

After locus analysis has been performed, the new joint reduces the same DOFs as the other joints between the two geoms together. Then, action analysis is performed on the new joint, which determines the relative DOFs left between the two geoms.

Underconstrained situations are repaired by adding constraints after the relative DOFs have been detected. When the relative DOFs of all pairs of geoms have been deduced, they are removed per pair by adding and solving constraints that reduce the DOFs between them. The DOFs have to be removed per pair of geoms, because otherwise an overconstrained situation could be introduced, since the relative DOFs of two different pairs of geoms can be the result of the same joint not reducing these DOFs. When the DOFs of this joint are reduced independently for multiple pairs of geoms, it could be done inconsistently.

An example of this is given in Fig. 10. Four groups of geoms exists whose relative position and orientation have not been completely fixed. When the relative DOFs of group A with respect to group C, and of group B with respect to group D, have been determined, they both contain a DOF that results from the joint between group B and group C. When this DOF is reduced by adding a constraint between group A and group C and between group B and group D at the same time, an overconstrained situation is introduced into the model.

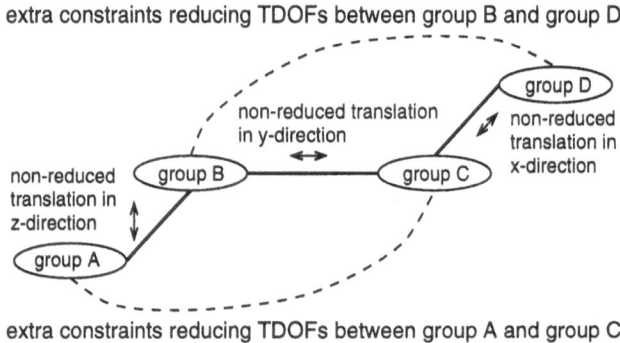

Fig. 10: Introducing an overconstrained situation while repairing an underconstrained one.

Currently, the constraints reducing the found DOFs are added by the user. The found DOFs could also be removed automatically using default constraints, as suggested by Suzuki et al. [SAK90].

Results and conclusions

To illustrate the approaches to solve over- and underconstrained models, an example is given here. Fig. 11 shows a model that consists of three blocks on the right hand forming a staircase, a horizontal block in the middle forming a bridge, and a block on the left hand that acts as a pier for the bridge.

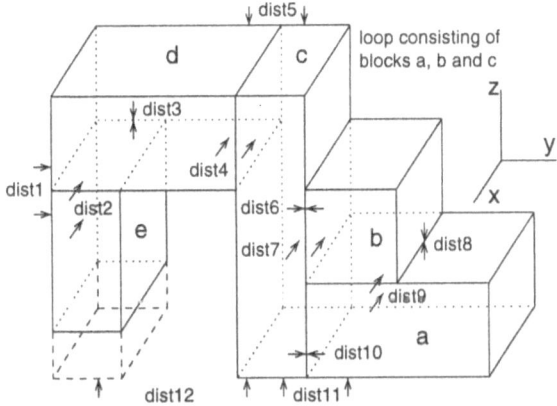

Fig. 11: A model with under- and overconstrained situations.

The model contains a loop that consists of the blocks of the staircase, because they are not rigidly connected. The model further contains an underconstrained situation, because the bridge and its pier are not completely fixed with respect to the staircase. Finally, the model contains an overconstrained situation, because the height of the bridge part is not equal to the height of block c of the staircase.

The model is solved by first merging block d and block e into part ed, because they have a fixed joint between them. After that, the model contains a rigid loop consisting of block a, block b and block c. This loop is solved by taking block c fixed and then rewriting the dist8 and dist9 constraints from the joint between block a and block b to the other joints of the loop, making them rigid. As a result of this, the blocks of the loop are merged into part abc, see Fig. 12.

After this, the model consists of two parts that are connected by an under- and overconstrained joint, because there is no constraint between them reducing translation in the y-direction, whereas constraint dist5 and constraint dist12 inconsistently reduce the translation in the z-direction between them. The constraints involved in this overconstrained situation can be found in the dependency graphs of the parts, see Fig. 13. Because the dist5 constraint works on block d and the dist12 constraint works on block e of part ed (see Fig. 11), the constraint dist3, between block d and block e of part ed, is involved. However, because the dist5 and the dist12 constraints both work on

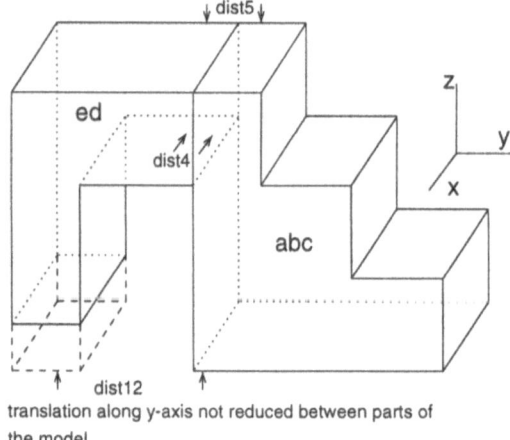

Fig. 12: The model of Fig. 11 after blocks with rigid joints have been merged,
and loops have been solved.

block c of part abc, no constraint from the dependency graph of part abc
is involved.

Let us assume that constraint dist12 has been chosen and removed,
so that the overconstrained situation has been repaired. After that, the
underconstrained DOFs are determined by taking one of the parts fixed,

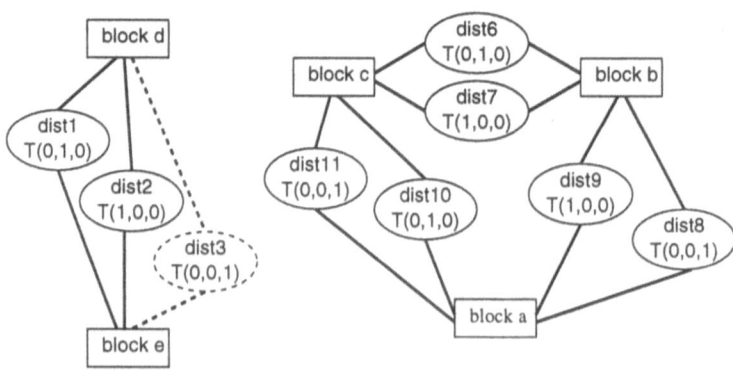

Fig. 13: The expanded dependency graphs of the blocks of the model of Fig. 11
when the overconstrained situation is found.

the other free, and then performing action analysis on the constraints of the joint between them. The DOFs left of the free geom after the constraints have been solved are the underconstrained DOFs. Then constraint dist13 is added that reduces the underconstrained DOFs, and the last two parts can be merged into part abcde, see Fig. 14.

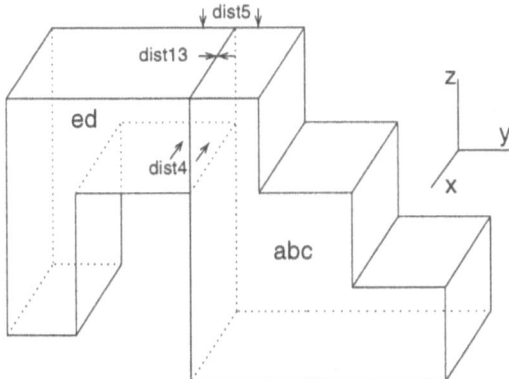

Fig. 14: The model of Fig. 11 after over- and underconstrained situations have been repaired.

A new approach has been described to solve over- and underconstrained models. The approach has been implemented in the geometric constraint solver that is used in the multiple-view feature modeling system SPIFF [BBDHK97], but it can be used with all non-iterative constraint solvers. It can deal with any shape supported by the modeling system. Future extensions to the presented approach include the following.

When a constraint is added to an underconstrained model to repair an underconstrained situation, this constraint should only reduce the underconstrained DOFs, because otherwise an overconstrained situation will be introduced in the model. To prevent the latter, the number of DOFs reduced by the constraints has to be taken into account by the system.

In case a constraint is removed from an overconstrained model, it should be done in such a way that no underconstrained situation is introduced. To achieve this, it should be tested whether the constraint to be removed only reduces the overconstrained DOFs, or also reduces other ones. If it does, the constraint should be replaced by a constraint that only reduces the other DOFs, so no underconstrained situation is introduced.

The system could be further improved by enabling it to automatically choose a constraint to be removed, in case of an overconstrained situation. The selection of the constraint could be done using several different criteria, such as removing the constraint that reduces fewest DOFs, removing the constraint that causes the least work to re-solve the model, and adding priorities to constraints and removing the constraint with the lowest priority.

As a final conclusion it can be stated that the capability to solve over- and underconstrained models has been shown to considerably enhance the usability of our DOF constraint solver.

References

[BBDHK97]   Bronsvoort, W. F., Bidarra, R., Dohmen, M., van Holland, W., de Kraker, K. J.: Multiple-view feature modelling and conversion, in: W. Strasser, R. Klein, R. Rau (eds), Geometric Modeling: Theory and Practice - The State of the Art, Springer-Verlag, 1997, pp. 159-174

[HB97]   Hsu, C., Brüderlin, B.: A degree-of-freedom graph approach, in: W. Strasser, R. Klein, R. Rau (eds), Geometric Modeling: Theory and Practice - The State of the Art, Springer-Verlag, 1997, pp. 132-155

[KKP95]   Keirouz, W. T., Kramer, G. A., Pabon J.: Exploiting constraint dependency information for debugging and explanation, in: V. Saraswart, P. van Hentenryck (eds), Principles and Practice of Constraint Programming; The Newport Papers, MIT Press, 1995, pp. 183-196

[Kra92]   Kramer, G. A.: Solving Geometric Constraint System: a Case Study in Kinematics, MIT Press, 1992

[SAK90]   Suzuki, H., Ando H., Kimura, F.: Geometric constraints and reasoning for geometrical CAD systems, Computers & Graphics, Vol. 14, No. 2 (1990)

A Constraint-Based Shape Modeling System

U. Döring, P. Michalik, B. Brüderlin[1]

We describe a new approach for geometric shape design which his centered around a 2D and 3D geometric constraint solver. This approach enables so-called non-history based modeling providing more flexibility to the designer. The modeler combines geometric and topological constraints, direct manipulation, sketching and Boolean set operations.

Introduction

The direct manipulation paradigm is now commonly applied in 2D illustration and 3D computer animation applications but it is still rarely used in 3D Computer Aided Design. Nevertheless, the need for intuitive tools with a simple point and click paradigm arises also for CAD applications, especially in the conceptual design phase. While in illustration and animation applications the need for accuracy is often limited to visual appearance, in mechanical CAD, data have to be manipulated at least at floating point precision. This is one reason why, traditionally, exact coordinate input, geometric constructions, and procedural and parametric modeling have been used in mechanical design. While such a traditional approach yields accurate data, it means a lot of work and planning ahead. Changing the parametrisation of an already specified part often means redesigning the object from scratch. In conceptual design one would like to start with a rough sketch, capturing the design intent, and then refine the input with exact dimensions, as one goes along. Another major application of so-called non-history-based modeling arises when CAD data stemming from a reverse engineering process, or from a hand sketch, need to be modified. These models do not have a parametrisation scheme

[1]Technical University of Ilmenau PF 100565, D-98684 Ilmenau, Germany

defined. Also, geometry data exchanged between CAD systems do not necessarily carry parametric information. Moreover, non-history based modeling is much more flexible than parametric modeling because it lets users manipulate models in a more associative, rather than a prescribed way. The approach we take here, is to combine the methods for direct manipulation and snap-dragging with discretized input, constraint solving, together with Boolean set operations, and topological constraints, in a unified framework. Gossard et. al. [GZS88] have proposed a similar approach to conceptual design a decade ago. However, at that time, only numerical iteration methods were available for solving the geometric constraint problem. Only now that more powerful geometric constraint solving methods are available, this approach for conceptual modeling really becomes feasible.

Related Work

A few researchers investigated the use of 'direct manipulation devices' (e.g. the Brown University widget set [ZHR+93] or the SGI Explorer library [CSH+92], as well as the direct manipulation devices described in [Rap93] or [NiOl87]). A very powerful conceptual modeling system, based on the direct manipulation paradigm was described by Jansen and van Emmerik [Emm93]. Those applications do not handle constraints. The article [RaEm93] describes a widget set to interact with a 2D constraint system. Snap-dragging in 3D was proposed by Bier [Bie86]. In [SoBr91] we describe an integration between snap-dragging and 2D constraints as well as some 3D degree of freedom analysis. In a previous publication [Bru86] we describe 'TWEAK', a constraint-based direct manipulator toolset for shape manipulations.

In [FFD93] the use of datagloves for three-dimensional placing of parts in assemblies is described. In this section presented here, we extend these previous approaches, by integrating the snap-dragging approach with a new 3D constraint solver.

Much research in constraint solving has been conducted in recent years, mostly in the 2D domain [Ald88, BeRo93, Bor81, Bru87, Bru93, BFH+95, HsBr93, HsBr96a, Hsu96, LGL81, Owe91, Rol91, SeGo86, SoBr94, Sut63], but also some 3D [Bru86, EHEB97, HsBr96b, HAH+97, Ros86, SoBr91, HoVe94]. A survey on constraint applications in CAD can be found in [AnMe95].

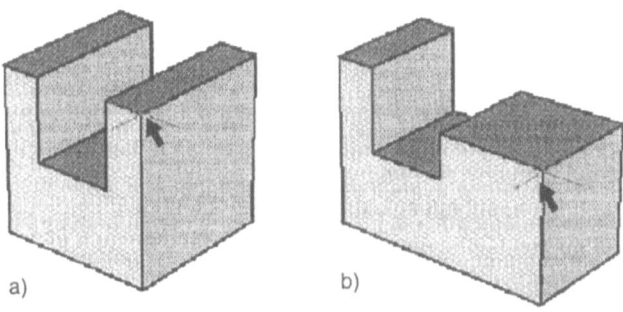

Figure 1: a) Picking a point to drag and b) dragging the picked vertex in three directions – adjacent faces follow because of constraints.

Geometric Constraints

It is well a known fact that interactive manipulations may violate the previously established data consistency of geometric objects. For example, although a polyhedral data structure implies that all vertices within the same plane are to be coplanar, one may inadvertently pull a point out of that plane, if the object is represented by a so-called b-rep structure. Other representations ensure consistency, at the cost of flexibility. For instance, a CSG representation always guarantees consistent solid objects, but it requires these objects to be defined by Boolean set operations, rather than by sweep operations, wireframes, or from hand sketches. History-based definitions define objects by a sequence of geometric constructions of any sort. However, this restricts the possible variations to changing the parameter values of the chosen constructions. Other modifications may require complete redefinition of the design, from scratch.

A constraint solver can be employed to maintain the consistency of shapes without the restriction in flexibility, known from other methods. In figure 1(a) we show the example of a box with a slot. At this point we impose no explicit constraints (e.g. right angles between adjacent planes), however, the coplanarity of the vertices in each face is imposed automatically, by the constraint solver integrated into the system. When picking the front right point, shown in figure 1(a), we can move it freely in all three directions. To ensure the planarity of the three faces, the constraint system will decide to translate the faces adjacent to the picked point (the result is shown in figure 1(b).

In a highly underconstrained object, such as in the example presented, there are infinitely many solutions to a posed constraint problem. Although there is no absolute best answer to such a problem, we expect that an interactive constraint system behaves predictably, when manipulating the selected elements. We apply two main principles to achieve this goal: typed constraints, and temporary constraints.

We distinguish between different types of degrees of freedom (i.e. translational, rotational and dimensional degrees of freedom) and apply a preference order when choosing the possible degrees of freedom in our analysis part of the solution. For instance, it seems obvious that translating a plane has a more predictable outcome than rotating it (rotating can be done about many different axes). Therefore, the translational degrees of freedoms are exploited whenever possible, before other degrees of freedoms are tried. A dimensional degree of freedom (e.g. the distance between two parallel planes) is also preferred over a rotational degree of freedom but less than a translational degree of freedom. In the example of the slot feature, this means that translating one side plane would also translate the opposite side (keeping the width of the slot constant, even without imposing a parallel distance constraint). If, however, a vertex of the opposite face has been fixed, then translating a vertex of the other side would change the distance between the parallel faces, although a rotation about the point is also a possible solution. Only after explicitly fixing the distance between the faces will a rotation be tried. In [EHEB97] we experimented with a mode-dependent preference order for the different types of constraints (by introducing so-called soft constraints). Soft constraints allow for even more flexibility, however having different modes can also be confusing.

Temporary constraints are used to interactively guide the constraint solver to choose among the available degrees of freedom. Each manipulation device (which is placed after picking a geometric element) constitutes a local constraint, as long as the handle is not actively manipulated. Local constraints guarantee that object manipulations cannot be inadvertantly undone by subsequent manipulations of other elements. As long as a handle is not explicitly deleted (which is simply done by double clicking on the handle) it remains for the rest of the editing session, and fixes all degrees of freedoms of that element. The handle can still be reactivated by dragging it, later on (while the local constraints are suspended). The temporary constraints are used to locally control the behavior of the constraint solver in a much more predictable way than a moded constraint hierarchy. The temporary constraints force the constraint solver to choose a different degree of freedom than it would choose, by default. We show a few examples, below.

In figure 2 we show how the example from figure 1 is further edited.

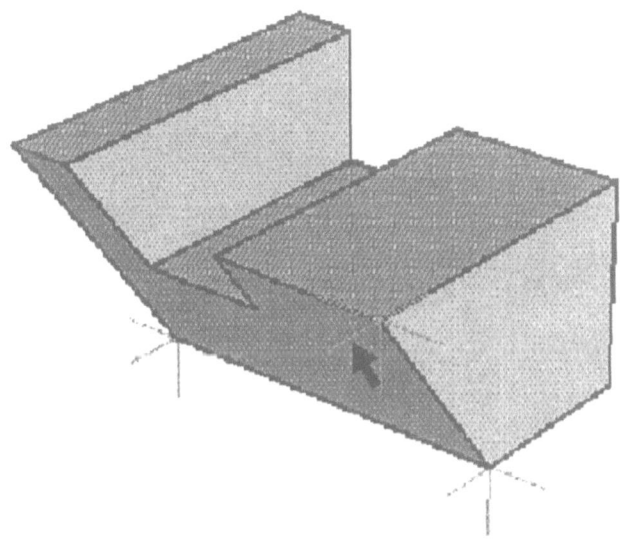

Figure 2: To constrain the rotation, pick two points on axis, and drag point forward.

Here, we want to rotate the front face about the bottom edge, while further dragging the same vertex as before. For this purpose, we fix the two vertices at the bottom of the face by picking each, and then dragging the top right vertex, again. The two vertex handles at the bottom act as temporary local constraints and force the constraint solver to rotate the plane around the axis defined by the two points at the bottom. In figure 3, we also fix two vertices at the back side of the object and then drag the top right handle upwards. The values displayed indicate that the dragged vertex was moved by 2.4 units in y, and by 2.0 units in z. It is also possible to rotate a face by a specified angle, about the same axis. Instead of picking and dragging a point in the face, we would pick the face and change its orientation by a specified angle in one direction.

Figures 4 - 5 show what happens when a face in the rectangular slot feature is rotated about one or two axes, using the face manipulation handle. The faces are automatically re-intersected to determine the new vertex locations. The result in figure 5 is a good example of an object which could no longer be defined by a simple sweep of a 2D profile. Without the constraint solver provided here, we would basically be forced to redefine the object by Boolean set operations. Using geometric constraints, the modification becomes simple and intuitive.

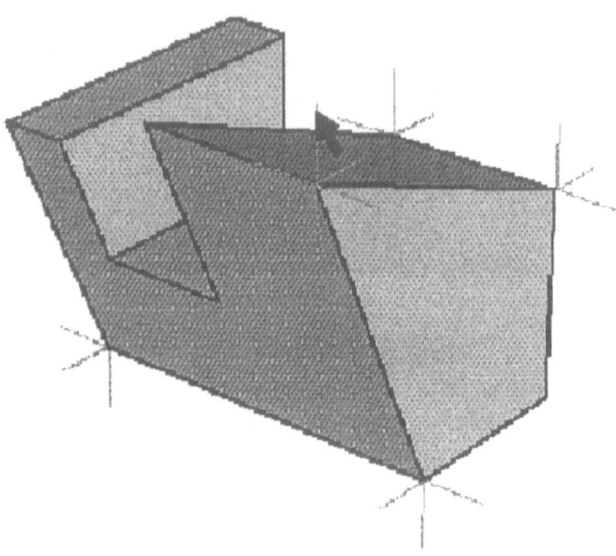

Figure 3: Constrain rotation axis on top face as well, and drag point up.

Graph representation of constraints

The geometric constraints, in our approach, are represented by a graph data structure. In this data structure, each geometric element (face, edge, vertex) is represented by a node, and a relationship (constraint) between two or more elements is represented by an arc in the graph (see figure 6). Each geometric element (node) has certain intrinsic degrees of freedom. A vertex, for instance, has three degrees of freedom, which are usually represented by the x/y/z coordinates. A plane also has three degrees of freedom (two angular degrees of freedom, representing the orientation of the plane, and one translational degree of freedom, representing the position). A geometric relationship (constraint) eliminates degrees of freedom between the elements connected by the constraint. The amount of reduction of degrees of freedom by a constraint is called the valency of the constraint [Ald88]. For instance, a parallelism constraint between two planes eliminates 2 degrees of freedom between the planes (i.e. two parallel planes have a total of 4 degrees of freedom as opposed to 6 for two non-parallel planes – the position and two orientation degrees of one plane, and the distance to the second plane). Therefore, a 'parallel' constraint between planes has a valency of 2.

A polygon in the so-called boundary representation can also be repre-

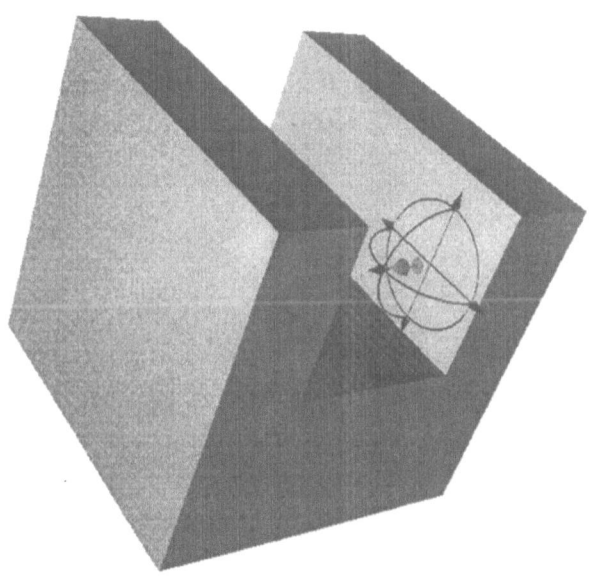

Figure 4: Picking a side of a rectangular slot feature.

sented as a constraint network. Each vertex in a face is related to that face by an incidence relation. A vertex-plane incidence is a valency 1 constraint (instead of 3 degrees of freedoms, the vertex is limited to the 2 dimensional plane, or vice versa, the plane can still be rotated in two directions about the point but loses its translational degree of freedom). By means of the incidence relationship, each vertex is incident on at least three planes (see figure 7), and all the vertices of the same face are constrained to be coplanar. Other constraints supported by the system are parallel distance (2 planes are parallel with a fixed distance: valency = 3), fixed angle between two planes (valency = 1) and fixed distance between 2 points (valency = 1).

Evaluating the constraint definitions in interactive situations

When selecting a geometric element for dragging, this action is interpreted by the constraint solver as requesting a certain numer of degrees of freedom from this element in the constraint network. For instance,

Figure 5: After orienting the slot feature.

repositioning a point in space means to request three degrees of freedom from that point; translating a plane means to request one degree of freedom, etc. In [HsBr96b] we describe how such a request can be met by the constraints, by propagating the request through the constraint network. That paper describes an efficient degree-of-freedom analysis algorithm which traverses the graph and finds a minimal subgraph of geometric elements that need to be updated as a consequence of the intended manipulation (by recursively requesting degrees of freedom from connected elements). The algorithm is based on the principle of conservation of degrees of freedom, which is expressed by the so-called balance equation for degrees of freedom for each node and arc. The result of the degree of freedom analysis is a dependency graph. The dependency graph is a directed version of a subset of the original constraint graph. The element from which we request degrees of freedoms becomes the root of the dependency graph. The dependency graph is then evaluated by a sequence of construction operations that construct the modified objects, in a top-down order, starting at the root. Those operations are triggered by so-called firing rules. Due to lack of space, we cannot give a detailed description of the degree of freedom analysis and the evaluation by construction oper-

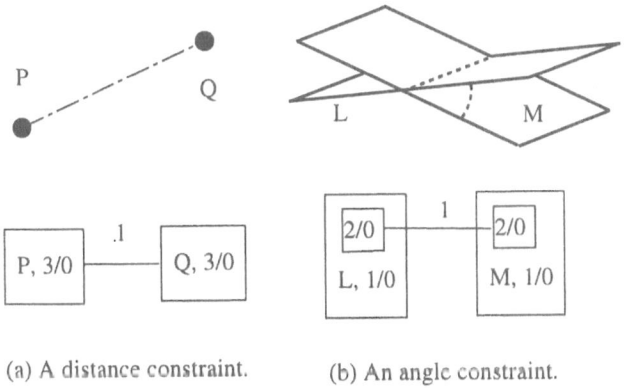

(a) A distance constraint.   (b) An angle constraint.

Figure 6: Constraining the distance between two points and the angle between planes.

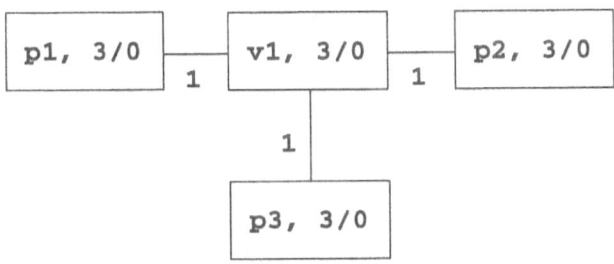

Figure 7: A constraint network representing the incidence relations between a vertex and three planes.

ations, here. Instead, we sketch the procedure by examples, below, and refer to [HsBr96b] and [HsBr96a] for details.

One of the main advantages of the degree-of-fredom-graph representation of the constraint problem, as well as the degree of freedom analysis algorithm, is that the problem is stated in a very abstract way, independent of the geometric interpretation and the dimension of the embedding space. We can therefore define problems in two and three dimensions, even mixed with algebraic equations between parameters (so-called engineering equations) with one unified framework. On the other hand, the required geometric constructions and algebraic operations necessary to evaluate the dependency graphs mean the implementation of a large number of functions with this approach. In principle, all operations typically occurring in CAD will be used. In the following, we describe some examples of applications.

Figure 8 shows a 2D triangle with variable distances. Two sides are congruent (length α), and the length β of the third side is related to the other two by the equation $\alpha + \beta = $ const (γ). These relations are represented by the graph in 8(b). The lengths themselves are represented by one-dimensional parameter objects. The algebraic equation is represented by the '+' relation which is a hyper arc (with valency 1) between the parameters. When requesting two degrees of freedom from point B, a dependency graph as shown in figure 8(c) is derived. The geometric evaluation is as follows: Point A may remain where it is, and point B follows the cursor. Then α is evaluated as the new distance between points A and B. Then β becomes the difference between γ and the new value of α. Then point C is constructed by intersecting two circles with radii α and β, respectively. The example shows how algebraic operations as well as geometric constructions are used in the evaluation.

Figure 9 shows an example of editing a polyhedron. All vertices are shared by several faces. In the constraint graph this fact is represented by incidence relationships between the corresponding planes and points. When requesting one degree of freedom from v1, the dependency graph shown in figure 9(b) is derived. The geometric interpretation is indicated under (a), as follows: Vertices v6 and v4 remain constant, plane p3 therefore rotates around the axis defined by the two points. The evaluation is again top-down, using geometric constructions. First, the new position of v1 (the root node) is determined by projecting the cursor position onto the line defined by intersecting planes p1 and p2 (which remain constant as well). Then p3 is recalculated from v4, v6 and the new v1. Afterwards v2 and v3 are recalculated by intersecting the new p3 with p5 and p6 (which also remain constant). All other vertices that belong to the top plane p3 also have to be recalculated by corresponding intersections (this is not shown on the dependency graph, in order not to clutter the illustration). All other parts of the polygon remain constant.

For polyhedra, the constructions are typically finding a vertex position by intersecting three planes, or finding a plane through three points, etc. With additional point - point distance constraints, the behavior is, of course, different, due to the additional constraints. Here the calculations also involve intersecting spheres with spheres and planes.

More on the evaluation of constraints by geometric construction operations can be found in the publication [HsBr96a]. The examples shown here can be updated at interactive speed on a low-end workstation or PC. We also ran examples involving several hundred constraints at interactive speed [EHEB97], and in most practical applications, unless a high-end graphics workstation is used, the rendering often seems to be the bottleneck, and not the constraint solving.

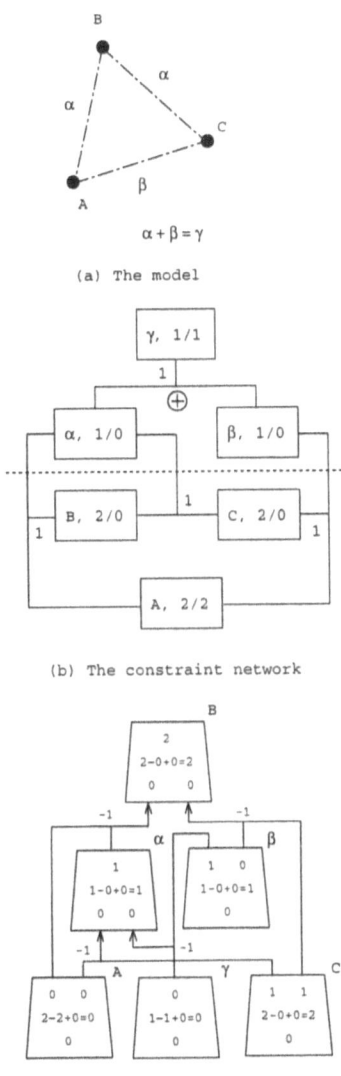

Figure 8: Mixed algebraic and geometric constraints. (a) A geometric model. (b) The constraint graph (c) A dependency graph for dragging point B.

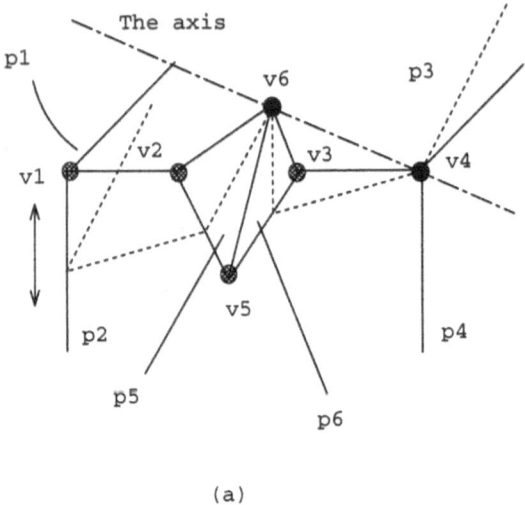

(a)

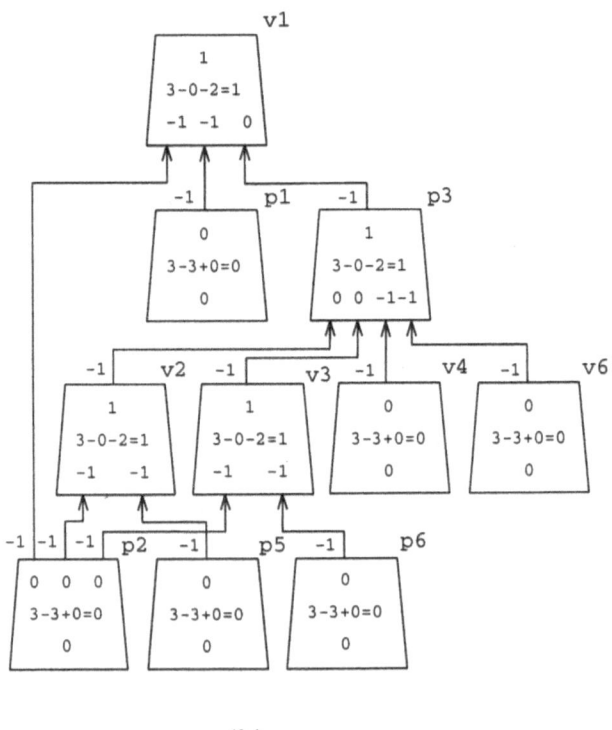

(b)

Figure 9: Editing a 3D polyhedron. (a) Moving vertex v1 along an edge (b) The dependency graph for dragging point v1.

Ensuring Topological Consistency

In our previous examples, we showed how the constraint definition ensures the geometric consistency, which was not necessarily guaranteed by the underlying representations alone. However, it is still possible to make objects inconsistent, as the following examples show. By dragging the selected vertices or planes too far, the object may become self-intersecting (see figure 10).

Our approach provides the following ways of ensuring topological consistency. Boolean set operations can be used to eliminate the inconsistencies. So, called inequality constraints, and topological constraints are used to avoid topological inconsistencies.

Topological tests are also built into the evaluation operations for each geometric constructor. For instance, if the solver decides to construct a point by intersecting two circles, there are generally two solutions. The default solution chosen is the one that corresponds to the one chosen in previous evaluation (for the example of circle-circle intersection, the solutions are distinguished by the orientation with respect to the line connecting the two circle centers). The user may also request alternative solutions, interactively. If the two circles are too far apart, they will not intersect, and the evaluation is interrupted and the previous solution is displayed. In some sense the limited domain of a constructor is like an inequality constraint, limiting the degrees of freedoms to some domain.

The example in figure 11 shows three points connected by distance constraints. One may drag point C with two degrees of freedom in the plane, without moving point A. However, the domain in which point C has two degrees of freedoms is limited to the shaded area in figure 12. Our solver also allows the explicit definition of inequality constraints. For example, we can define the distance between point A and C to be between 3 and 5cm (figure 13). The resulting range of point C is exactly the same as in figure 12.

Depending on the user preference, in the case of violating an inequality constraint, the degree of freedom analysis algorithm may also be reinvoked, to request degrees of freedoms form other parts of the model. For the example of figure 11, this means if we drag point C outside the domain, we have to move point A as well. This interactive behavior can be selected as an option by the user.

In addition to distance inequalities, angle inequality constraints can be expressed in our system. Such inequality constraints are useful for the definitions of mechanical devices, for instance, to limit the range of motion.

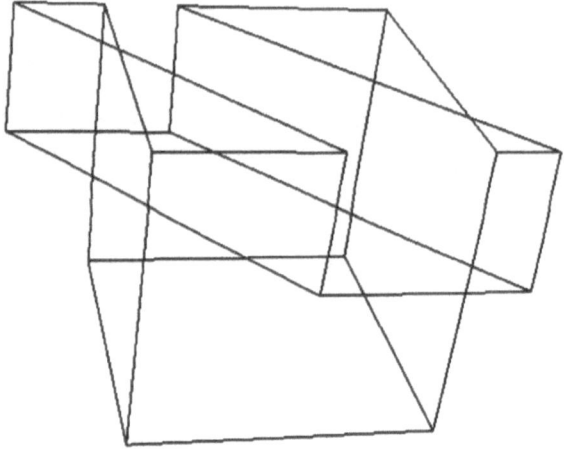

Figure 10: After dragging a face too far, a topologically inconsistent configuration occurs.

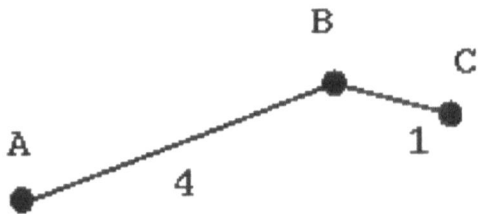

Figure 11: Three points with distance constraints.

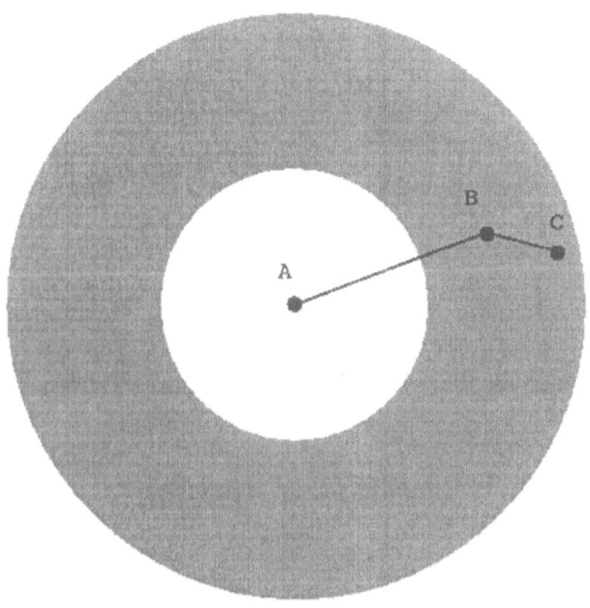

Figure 12: Range of point C with respect to point A.

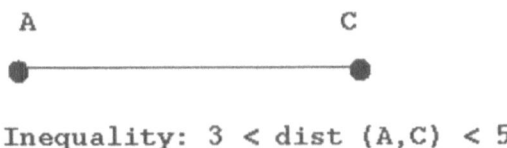

Figure 13: Expressing the distance between A and C as inequality.

In many cases topological constraints such as non-self-intersection are important, as shown in example.

Boolean set operations are used to eliminate inconsistencies, due to self intersection of objects (e.g. as shown in figure 10). For this purpose, we use our own implementation of Boolean set operations, which is described in [Bru90, Bru91, Fan92, ZFB93].

If an object is already defined by set operations, it suffices to reevaluate the Boolean operation after each geometric modification. In the current implementation we assume that we have the object defined by a Boolean expression over half spaces. The set operation algorithm finds a boundary representation, consisting of facets, loops, edges, and vertices from this expression. The half spaces refer to the same planes that are used in the constraint network. The new vertices are also put in the constraint network, and the incidence relations between faces and vertices are automatically established after the evaluation of the set operation. Manipulating faces or vertices with the constraint solver therefore automatically updates the dependent planes in the Boolean expression, and reevaluating the expression will therefore yield a topologically consistent object, again (see figure 14).

If no Boolean expression is available (which is the case if the object originally was created by a sweep operation, or a hand-drawn sketch, etc.) we will need some special shape operator to eliminate self-intersecting parts of the object, represented in boundary representation.

Instead of eliminating the inconsistency after it occurs, we may avoid it by detecting the self-intersection, and disallow the solution. We then eliminate the corresponding degree of freedom (as it is done when an inequality constraint is violated), and we don't let the user drag further in the same direction. Avoiding topological inconsistencies is preferred when the functionality of features depends on topological relationships.

Conclusion

The added flexibility offered by using a constraint solver has the promise of significantly shortening the design process. Even the simple examples shown in this section would require much more complex definitions or repeated redesign, to accomodate the new design decisions, in current CAD systems. The editing operations proposed in our approach can be used, for instance, for editing objects that come from a sketch input (such as described in [EHEB97]) but also in combination with conventional CAD

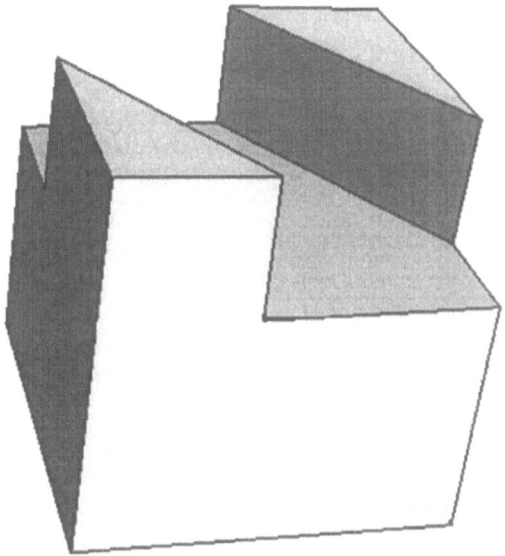

Figure 14: Recalculating the Boolean set operation eliminates self intersections.

operations (e.g. sweep or CSG) or to edit objects generated by reverse engineering.

It becomes clear that the added flexibility which a constraint-based approach provides over a parametric approach demands new modeling paradigms. In a parametric definition, changing a parameter value, in most cases, has an unambiguous outcome. Since geometric constraints are bi-directional, there are many different ways of interpreting such a change, especially in underconstrained situations. In this section we described a few concepts to disambiguate the many choices. This is done by temporary local constraints (which give more control to the user) a preference order between degrees of freedoms, and by topological constraints, limiting the range of motion. We also combine the constraint solver with Boolean set operations to eliminate self-intersecting parts of the models, after they occur.

The modeler described in this section is a first step in the direction of non-history-based modeling. We need to gain more experience with this approach which will hopefully lead to more powerful modeling paradigms in the future.

References

[Ald88]   B. Aldefeld. Variation of geometries based on ageometric-reasoning method. *Computer Aided Design*, 20(3):117–126, April 1988.

[AnMe95] R. Anderl and R. Mendgen. Parametric design and its impact on solid modeling applications. In *Proceedings of the Third Symposium on Solid Modeling and Applications*, Salt Lake City, 1995. ACM Press.

[BeRo93]   R. Berling and M. Rosendahl. Geometry modeling using dimensional constraints. In *9th International Conference on CAD/CAM, Robotics and Factories of the Future*, 1993.

[Bie86]   E. Bier. Snap dragging in 3d. In *Proceedings of the 1986 Workshop on Interactive 3D Graphics, ACM SIGGRAPH*, Chapel Hill, North Carolina, 1986.

[Bor81]   A. H. Borning. The programming language aspects of ThingLab, a constraint-oriented simulation laboratory. *ACM*

Transactions on Programming Languages and Systems, 3(4):353–387, October 1981.

[BFH+95] W. Bouma, I. Fudos, C.M. Hoffmann, Jiazhen Cai, and Robert Paige. A geometric constraint solver. *Computer Aided Design*, 27(6):487–501, June 1995.

[Bru86] B. Bruderlin. Constructing three-dimensional geometric object defined by constraints. In *Proceedings of the 1986 Workshop on Interactive 3D Graphics, ACM SIGGRAPH*, Chapel Hill, North Carolina, 1986.

[Bru87] B. Bruderlin. *Rule-Based Geometric Modelling*. PhD thesis, ETH Zürich, Switzerland, 1987.

[Bru90] Beat Bruderlin. Detecting ambiguities: An optimistic approach to robustness problems in computational geometry. Tech. Rep. UUCS 90-003, Computer Science Department, University of Utah, April 1990.

[Bru91] Beat Bruderlin. Robust regularized set operations on polyhedra. In *Proc. of Hawaii International Conference on System Science*, January 1991.

[Bru93] B.D. Bruderlin. Using geometric rewrite rules for solving geometric problems symbolically. *Theoretical Computer Science*, 2(116):291–303, August 1993.

[CSH+92] D. Brookshire Conner, Scott S. Snibbe, Kenneth P. Herndon, Daniel C. Robbins, Robert C. Zeleznik, and Andries van Dam. Three-dimensional widgets. In David Zeltzer, editor, *Computer Graphics (1992 Symposium on Interactive 3D Graphics)*, volume 25, pages 183–188, March 1992.

[EHEB97] L. Eggli, C. Hsu, G. Elber, and B. Bruderlin. Inferring 3d models from freehand sketches and constraints. *Computer Aided Design*, February 1997.

[Emm93] M.J.G.M. van Emmerik. *Interactive design of parameterized 3D models by direct manipulation*. PhD thesis, Delft University, 1990.

[FFD93] M. Fa, T. Fernando, and P.M. Dew. Interactive constraint-based solid modelling using allowable motion. In *Proceedings of the 1993 ACM/SIGGRAPH Symposium on Solid Modeling Foundations and CAD/CAM Applications*, Montreal, Canada, May 19-21 1993.

[Fan92]   S.F. Fang. *Robustness In Geometric Modeling*. PhD thesis, University of Utah, 1992.

[GZS88]   D. Gossard, R. Zuffante and H. Sakurai. Representing Dimensions, Tolerances and Features in MCAE Systems. In *IEEE Computer Graphics & Applications*. March 1988.

[HoVe94]   C. Hoffmann and P.J. Vermeer. Geometric constraint solving in R^2 and R^3. In *Computing in Euclidean Geometry, 2nd Edition*. World Scientific Publishing Co Pte Ltd, 1994.

[Hsu96]   C. Hsu. *Graph-Based Approach for Solving Constraint Problems*. PhD thesis, Computer Science Department, University of Utah, May 1996.

[HAH+97]   C. Hsu, G. Alt, Z. Huang, E. Beier, and B. Brüderlin. A constraint-based manipulator toolset for editing 3d objects. In *Proceedings of the 1997 ACM/SIGGRAPH Symposium on Solid Modeling Foundations and CAD/CAM Applications*, Atlanta Georgia, 1997.

[HsBr93]   C. Hsu and B. Brüderlin. Constraint objects - integrating constraint definition and graphical interaction. In *Proceedings of the 1993 ACM/SIGGRAPH Symposium on Solid Modeling Foundations and CAD/CAM Applications*, Montreal, Canada, May 19-21 1993.

[HsBr96a] C. Hsu and B. Brüderlin. A hybrid geometric constraint solver using exact and iterative geometric constructions. *Presented at the Dagstuhl Seminar "CAD-Tools for Products", September 1995, in "CAD Tools and Methods for Design System Development", D. Roller and P. Brunet eds, Springer-Verlag*, 1997.

[HsBr96b] C. Hsu and B. Brüderlin. A graph-based degree of freedom analysis algorithm to solve geometric constraint problems. In *Proceedings of Theory and Practice of Geometric Modeling (Blaubeuren II)*, Blaubeuren, October 1996 (Can be found in *Geometric Modeling: Theory and Practice*, W. Strasser and R. Klein eds, Springer-Verlag, 1997).

[LGL81]   V.C. Lin, D.C. Gossard, and R.A. Light. Variational geometry in computer-aided design. *Computer Graphics*, 15(3):171–177, August 1981.

[NiOl87]   G. Nielson and D. Olsen, Jr. Direct manipulation techniques for 3D objects using 3D locator devices. In *Proceedings of the 1986 Symposium on Interactive 3D Graphics*, 1987.

[Owe91]   J.C Owen. Algebraic solution for geometry from dimensional constraints. In *Proceedings of the 1991 ACM / SIGGRAPH Symposium on Solid Modeling Foundations and CAD / CAM Applications*, May 1991.

[Rap93]   A. Rappoport. Direct manipulation devices for the design of geometric constraint networks. In Magnenat-Thalmann N. and Thalmann D., editors, *Communicating with Virtual Worlds – Proceedings Computer Graphics International, Lausanne*. Springer Verlag, 1993.

[RaEm93] A. Rappoport and M.J.G.M van Emmerik. User interface devices for rapid and exact number specification. *ACM Transactions on Graphics*, 12:348–354, 1993.

[Rol91]   D. Roller. An approach to computer-aided parametric design. *Computer Aided Design*, 23(5):385–391, June 1991.

[Ros86]   Jaroslaw P. Rossignac. Constraints in constructive solid geometry. In *Proceedings of Workshop on Interactive 3D Graphics*, pages 93–110, Chapel Hill, NC, October 23-24 1986.

[SeGo86] D. Serrano and D.C. Gossard. Combining mathematical models and geometric models in CAE systems. In *Proc. ASME Computers in Eng. Conf.*, pages 277–284, Chicago, July 1986.

[SoBr91] W. Sohrt and B.D. Brüderlin. Interaction with constraints in 3D modeling. *International Journal of Computational Geometry and Application*, 1(4):405–425, December 1991.

[SoBr94] L. Solano and P. Brunet. Constructive constraint-based model for parametric cad systems. *Computer Aided Design*, 26(8), 1994.

[Sut63]   I. Sutherland. *Sketchpad, a man-machine graphical communication system*. PhD thesis, MIT, January 1963.

[ZHR+93] Robert C. Zeleznik, Kenneth P. Herndon, Daniel C. Robbins, Nate Huang, Tom Meyer, Noah Parker, and John F. Hughes. An interactive 3D toolkit for constructing 3D widgets. In James T. Kajiya, editor, *Computer Graphics (SIGGRAPH '93 Proceedings)*, volume 27, pages 81–84, August 1993.

[ZFB93]   Xiaohong Zhu, Shiaofen Fang, and Beat Bruderlin. Obtaining robust boolean set operation for manifold solids by avoiding and eliminating redundancy. In *Proc. of Second Symposium on Solid Modeling and Applications*, May 1993.

Chapter 3

Constraint Representation and Solving Methods

Overview

Modelling of Geometric Constraints in CAD-Applications

Manfred Rosendahl[1] , Roland Berling[2]

Introduction

Modelling in CAD-applications demands a mapping of the structure of the geometric objects and a description of the relation between them. In purely algebraic modelling there are equations between the elements of the geometric model defined, which are solved to find the characteristic points of the model[SeGo87]. Another way is to use geometric reasoning mechanism in which the dimensions and geometric relationships are defined as either facts or rules [RSV89,VSR92]. To model the relations constraint systems are well suited. In this paper we describe two methods.

The first approach, a type of *constructive* modelling, describes the geometric and non-geometric relations by construction operations, which will be used in building and modifying a geometry. The result is a mapping which describes the construction completely by a set of parameters. The modifying of dependent values is done either by reverse operations or by solving an equation system. Additional constraints can be defined on top of a given construction. Therefore constructions which need circular definitions can also be defined. This constructive approach, used in the **RelCAD** system, reduces the number of equations which have to be solved. Another feature which has been introduced to the constraint based geometric modelling is the Segment-

[1] Institut für Informatik, Univerität Koblenz, Germany
[2] Hewlett Packard, Böblingen, Germany

concept. In normal CAD-systems collections of elements, called group, block or segment are either programmed or the parameters can only be insertion point, angle, size. In the **RelCAD**-System a segment is a collection of objects that not only logically belong together, but also have geometric/numerical relationships among each other. They can even have relationships to objects outside the segment and can thereby be dependent on them. Any object of a segment definition can be treated as a parameter, not only dimensions or points but also other items and even segments [DRB93,RBD96].

In this approach the a model is developed in which the directions of the propagations are fixed. To evaluate a model against the direction in the constraint graph new constraints are added, which result in a graph with cycles. The second approach, a *declarative* modelling, describes the geometric relations by basis-constraints on points and dimensions as variables. The resulting constraint-graph will be evaluated by local propagation. The specific type of modelling makes it possible to use constructive operations, when evaluating the graph from any sufficient set of given variables. By this manner numerical methods like iterations can be avoided in most cases. This technique is developed out of the first approach and is suited for the early stages in a design process. This works especially because first a model with an undirected graph is developed, from which the free and derived variables can still be chosen. [BeRo93, Ber96].

Constructive Modelling: RelCAD Geometric modelling system

The **RelCAD** system was developed at the University of Koblenz as an application extension to the general 2D CAD-system **VarioCAD**, also developed there.

Each element of the geometric model is either an absolute one or defined by its relations to other geometric objects. So any changes in an object cause changes in all objects, which directly or indirectly depend on them. This is achieved because the dependencies built a directed a-cyclic graph (DAG). The method for changing an object, which is not a leaf in the graph, will be explained later.

The following objects are defined as absolute geometric and non-geometric objects:

Value (dimension), e.g. co-ordinate, length, distance, radius, angle but also non-geometric e.g. engineering values,
Point,
Line,
Circle,
Arc.

The relationships between the objects are typed according to these absolute object types. Only Arc is a subclass of Circle and can be used if a circle is required.

From these absolute classes, which have no relations the constructive classes with relations are derived. The relations are defined in all directions e.g. a point can be derived from geometric elements and also geometric elements can be derived from points.

Some examples of classes which are defined by the corresponding absolute class:

- Value:
  Co-ordinate, Distance, Angle, Expression
- Point:
  Intersection point, Tangent point, Centre point, Endpoint, Relative-point
- Line:
  Tangential line, Perpendicular Line, Parallel Line, Axis parallel line
- Circle:
  Given by centre and Radius, Tangential to 2 lines or circles and radius, Tangential to 3 lines or circles,
- Arcs:
  same arc circles

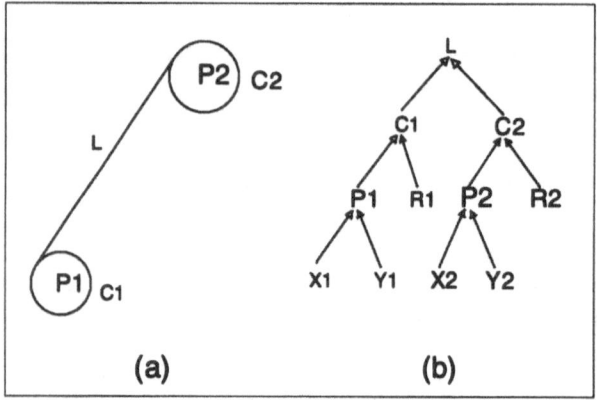

Figure 1

Figure 1 shows the definition of a derived element of the type Line_2objects. This element has 2 support elements C1 and C2.

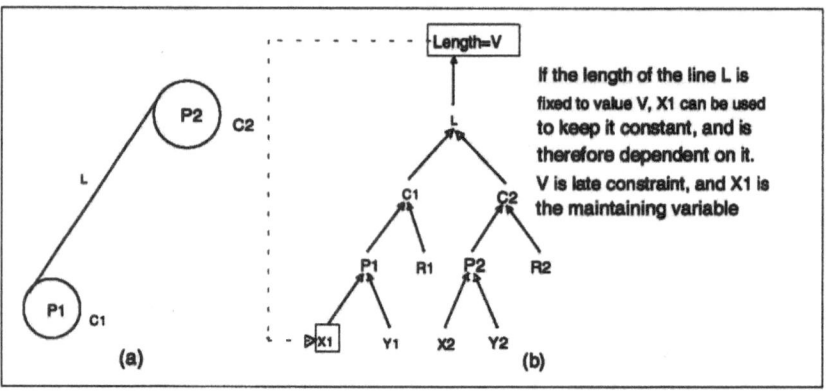

Figure 2

Up until now we have only discussed definitions with a-cyclic dependencies. There is also a class, which deals with cyclic dependencies. It is used to set extra dimensional restrictions on derived objects. For example, it could be used to fix the length of the line L in figure 2 This kind of constraint is called **late constraint,** because it is mostly assigned to the objects after they have been created. We call the derived object to which the late-constraint is assigned a **late-constrained-object**. A late constraint causes cyclic dependency among the geometric objects, because as an extra independent dimensional constraint it influences the original dimension of a late-constrained-object and thereby the support objects of it, which again determine the original dimension.

For example, if the length of the line L in figure 2(a) is fixed, the positions and radii of both circles $C1$ and $C2$ are also restricted. Changing of them leads to a irregular value for the length of the line L which does not agree with the fixed length.

Figure 2.(b) illustrates the situation of cyclic dependency. In order to maintain the consistency of the late constraint a free dimension which directly or indirectly supports the late-constrained-object should be selected as a maintaining variable. A maintaining variable of a late constraint has the following tasks:
1) When the value of the late constraint has changed the maintaining variable should also be changed so that one of the support objects of the late-constrained-object gets the new data which leads to a consistent late constraint.
2) When the support objects of the late-constrained-object have changed the maintaining variable should be changed so that the late constraint remains satisfied.

A dimension is no longer free if it is selected as the maintaining variable of a late constraint. Figure 2(b) shows an example where the co-ordinate X1 is selected to be the maintaining variable if the length of the line L is fixed to be the value V.

The evaluation of a late-constrained-object is more complicated than the evaluation of other geometric objects because of the cyclic dependency among the late-constrained objects, its late-constraint, and the maintaining variable of the late-constraint. An iterative procedure is used to find an appropriate value for the maintaining variable to satisfy the late-constraint. More than one late constraint can exist in a model at the same time and some of them may be related to each other when the maintaining variable of one late constraint is the support object of another late-constrained-object. The related late constraints should be satisfied simultaneously, which means we have to solve a set of (non-linear) constraint equations, where the unknowns are the maintaining variables. For each geometric model there is one process to satisfy all late constraints (independent and related) by using the Newton-Raphson iterative method. When evaluating a geometric model this process will not be run until all other objects of the model have been evaluated.

Because the late constraints are necessary only when the geometric model can not be constructed in sequential order any more and because they are assigned to the model after the main part of the model has been sequentially constructed, the number of the constraint equations which have to be solved simultaneously can be reduced by the user.

Late constraints are also used to change the data of conditioned objects, that means objects, which are not a leaf in the relation graph. If the user wants to change such an object, he has to choose, which independent objects(leafs) should be given free. Then a temporarily late constrained is solved to find a solution with the new values of the dependent objects.

Parametric design using segments

A family of design parts with various dimensions are often needed during the design. The concept segment serves this purpose.

About segments

The concept of the segment is similar to the concept of the *procedure* in high-level programming languages, *e.g.* Pascal. Therefore we introduce our segment by comparing it with the procedure concept.

The segment has two related aspects: segment scheme and segment instance. Segment scheme describes the inner structure of the segment. It determines what the components of the segments are, *e.g.* the number of the components, the type of each component and the relationships among the components. A segment instance is a graphical realisation of a segment schema. It is derived from the segment scheme and the actual parameters and is an instance and a graphical representation of the segment. Segment schemes correspond to *procedure declarations* and segment instances correspond to *procedure calls*.

In the following discussion we just use the word 'segment' when it is not necessary to differentiate the segment scheme from the segment instance. We say 'the components of the segment' when it is not necessary to know exactly how the components are defined within the segment scheme and the segment instance.

Another significant feature of the segment is that it can also have parameters. Similar to the parameters of the *procedure* there are two different forms of parameters for the segment: formal parameters and actual parameters. The formal parameters exist in the segment scheme and determine the types and the sequences of the actual parameters. The actual parameters do not belong to the segment instance, but are used by the segment instances to determine the size, the variational shape and the position of the segment instance. The actual parameters define the relations of the segment to the rest of the model.

Each *procedure* can be called many times in a program. Thus, a segment scheme can also be associated with many segment instances, which means that some segments may have different graphical representations. This can happen, when the data of the actual parameters are different for each segment instance with the same inner structure.

The class segment

The graphical representations of the segments are called 'segment instances' because they could be treated as the instances of the class 'segment scheme' from the point of view of the object-oriented methodology. In our approach we define these two concepts as separate classes and set up a connection between them.

Two new classes are defined. They are also called segment scheme and segment instance. A segment scheme contains
(a) a list of formal parameters, and
(b) a list of components.
The component objects are mostly related to each other and some of them have relationship with formal parameters.

A segment instance contains
(a) a list of actual parameters,
(b) a corresponding segment scheme
Segment scheme and segment instance are also treated as classes (or types) like the other classes. An instance of the segment scheme class is a concrete segment scheme with a definite number of geometric objects as components and a number of formal parameters of certain types. A concrete segment scheme does not appear in the geometric model. An instance of the segment instance class is the representation of a certain concrete segment scheme in the geometric model. Figure 2 shows the segment scheme class, segment instance class and their instances.

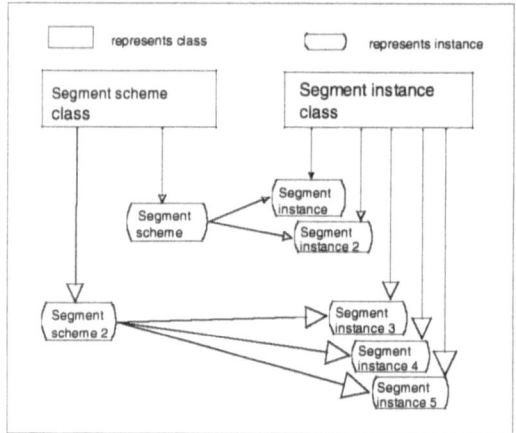

Figure 3

In the following discussion we call the instance of the segment scheme class **segment scheme** and the instance of the segment instance class **segment instance** when there is no misunderstanding.

The parameters of the segment

The formal parameters can be either
  (1) of any basic type, *e.g.* value, point, line, circle or arc.
  (2) of a segment instance.
Formal parameters are support objects of some component objects in a segment scheme. Therefore they determine the graphical data of these component objects.

An actual parameter must be of the same type with the corresponding formal parameter or a type derived from it, *i.e.* a object in the same absolute object-class. Therefore an actual parameter can be a derived object. Actual parameters do not belong to the segment instance. They are local geometric objects in the geometric model. If the formal parameter is a segment instance the actual parameter must be an instance of the same segment schema.

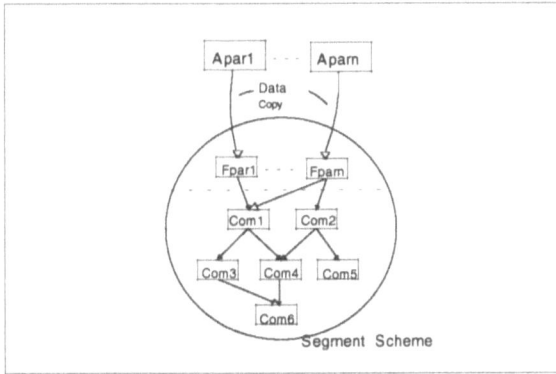

Figure 4

This is also the way compatible elements are defined.

Definition: 2 elements are compatible if they are either
- Derived from the same absolute object class
- Instances of the same object sheme

The data of the actual parameters determine the size, the variational shape and the position of a segment instance through formal parameters. The computing process of a segment instance does the following operations:

(a) It copies the graphical data of the actual parameters into the formal parameters of the associated segment scheme;

(b) It runs all the computing processes of the component objects in the associated segment scheme to generate the graphical data for each component;

(c) It returns all the graphical data obtained in (b) to the segment instance.

Figure 4 shows the actual and formal parameters and the segment scheme. The arrows within the segment scheme illustrate the supporting relation of component objects. With different actual parameters variations on the segment scheme can be generated. When a segment instance uses local objects of a model as actual parameters it is fully embedded in the geometric model.

The external object of the segment

Although the component objects of the segment are geometrically related to each other it is still possible for them to access the objects outside the segment, which means that a component object of a segment has objects outside the segment as its support objects. Such 'outside' object is called **external object**. This is equivalent to the global references in procedures. In procedures there is no method (but also no need) to access the local variables from outside. But in CAD accessing component objects of a segment from outside must be possible. An access to a component object of a segment is realised as an access to a substitute object of this component object. Also segments with alternatives and iterations are realised in the RelCAD system. For a more detailed description of the segment-concept see [RMD96].

Currently the concepts of the RelCAD system are transferred to a 3D system. Here new 3D oriented relations have to be defined, e.g. a cone defined by an apex point and a tangent sphere or a cone defined by two tangent spheres. Moreover a lot of 3D geometry is defined by 2D contours, e.g. a 3D object defined by the sweep of a 2D contour. Here the concepts of the 2D system can be used for defining the contour.

Declarative Modelling

In contrast to the constructive approach, the declarative approach describes the relations by a set of equations. Geometric relations between the lines, circles, etc., will be reduced to a few basis-relations between points and dimensions. The constraint model contains a set of basic geometric elements like:

- Segment(P1, P2): line segment from point P1 to P2

- Circle(M, r): circle centre point M and radius r

- Arc(P1, P2, M, r): arc from point P1 to P2 with centre M and radius r

and a set of constraint relations like:

- dp(P1, P2, d): distance between point P1 and P2 is d

- dl(P, Pa, Pe, d): distance between point P and line from Pa to Pe is d

- a3(P1, P2, P3, α): the angle between the line P1 to P2 and the line P1 to P3 is α

- a4(P1, P2, P3, P4, α): the angle between the line P1 to P2 and the line P3 to P4 is α

- equ(<expression>): the value of the expression is 0. If the expression contains n variables, it must be possible to compute one variable if the other n-1 variables are known.

- fix(<value>): a constraint without input values and just one output value.

Constraints between the geometric elements can be described by one or more of these basic constraints. Constraints with more than one degree of freedom are also described by a graph with these basic constraints.

This model in figure 5 can be derived by the following constraints:

dp(M1,T1,r1), dp(M2,T2,r2), dl(M1,T1,T2,r1), dl(M2,T2,T1,r2), circle(M1,r1), circle(m2,r2), line(T1,T2)

Non-geometric constraints can be described, too. For example, expressing that the length of the tangent-line is related to the sum of the radii by a factor f can be done by the following constraints:

dp(T1,T2,d), equ(d-(r1+r2)*f)

The constraint system is represented by an undirected graph $G(V,E,\varnothing)$ with nodes $V=P \cup D \cup C$ representing the points (P), the dimensions(D) and the constraints(C), and edges $E \subset C \times (P \cup D)$ representing the constraints bindings. There are 3 types of edges between a constraint and a variable.

none: undirected edge.

c_in: directed edge from the variable to the constraint.

v_in: directed edge from the constraint to the variable.

To define algorithms on the structure graph, we need operations to define the edges of certain types adjacent to a node. Therefore we define the following mappings:

Λ: $V \times$ typeId $\rightarrow$ P(E): set of edges of typeId adjacent to node V

Γ: $V \times$ typeId $\rightarrow$ P(V): set of nodes adjacent to V via edge of typeId

δ: $V \times$ typeId $\rightarrow$ N_0: number of edges of typeId adjacent to node V

With respect to the orientation of the edges the following mappings are defined:

Λ_{in}: V → P(E): set of oriented edges incoming to a variable or a constraint .

Λ_{out}: V → P(E): set of oriented edges outgoing from a variable or a constraint .

Λ_{none}: V → P(E): set of non-oriented edges adjacent to a variable or a constraint .

Analogously the mappings $\delta_{in}(v)$, $\delta_{out}(v)$ and $\delta_{none}(v)$ are defined as the number of edges adjacent to v with the equivalent predicate. The mappings $\Gamma_{in}(v)$, $\Gamma_{out}(v)$ and $\Gamma_{none}(v)$ are defined as the sets of nodes adjacent to v via an edge with the equivalent predicate.

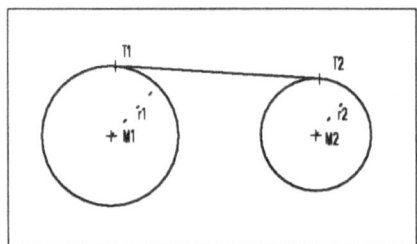

Figure 5

Figure 6 shows the graph for the example in figure 5.

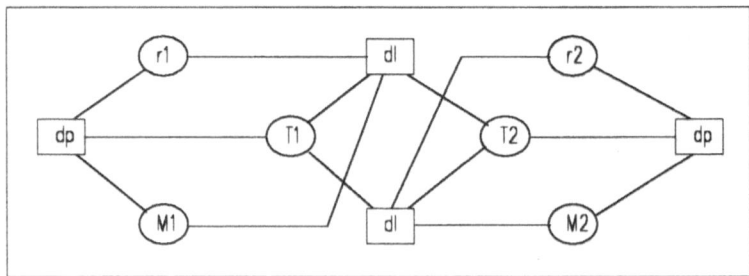

Figure 6

The undirected graph unveils the structure of the possible equation systems which can be formed by instantiating new values for parameter vertices or changing old ones. To fix a dimension or point, one respective two constraint fix with an edge to the fixed object will be added. By means of the constraint satisfaction process the undirected graph is transferred to a directed graph, which represents the evaluation of the geometry. This evaluation can be done by local propagation as long as there are no cycles in the directed graph, otherwise iterative methods have to be used. Therefore the goal of the orientation is to obtain as few cycles in the graph as possible.

The oriented graph has to fulfil the following restrictions.

$\forall \, c \in C: \delta_{out}(v) = 1$

$\forall \, v \in P: \delta_{in}(v) = 2$

$\forall \, v \in D: \delta_{in}(v) = 1$

Each point needs two values to be determined and each dimension one. These are the incoming edges. The values are then propagated through the outgoing edges. Each variable has a structural degree of freedom (sdf), which gives the degree of freedom, that is not yet restricted by constraints, i.e. by incoming edges.

$\forall \, v \in P: sdf(\{v\}) = 2 - \delta_{in}(v)$

$\forall \, v \in D: sdf(\{v\}) = 1 - \delta_{in}(v)$

$\forall \, M \subseteq P \cup D: sdf(M) = \sum_{v \in M} sdf(\{v\})$

In an undirected constraint graph $G(V, E, \varnothing)$, $V = C \cup P \cup D$ exists a complete matching of the set of constraint nodes C in the set of variable nodes $P \cup D$ if and only if to each constraint subset $N \subseteq C$ in the adjacent variable set there exist at least as many structural degrees of freedom as there are elements in N.

The constraint satisfaction process determines (in three phases) a complete allocation of the constraints to the variables. The allocation is defined by an orientation of the edges. The method identifies subgraphs which are structurally over-determined. To determine the allocation graph, algorithms to compute maximal matching [MH90] can be used. Serano and Gossard use this in their approach [SeGo87]. The method here uses propagation to get the smallest constraint graph on which a matching algorithm has to be applied. The allocation of a constraint to a variable will be expressed by the edge attribute v_in. All other edges adjacent to this constraint are then input values, which are attributed c_in.

A constraint graph is valid, if the following holds:

- Constraint node c: At most one v_in edge. If there is one v_in edge, all other edges are oriented.

- Variable node v: At most as many v_in edges as there are degree of freedom. If there are exactly as many (sdf(v)=0), then all edges are oriented.

During the propagation process the graph has to be always valid. The three phase of the method are:

- Propagating the constraints

- Propagating the degrees of freedom

- Determining the maximal constraint matching

The first two steps are used to minimise the constraint set, which has be applied to step 3. The propagating of constraints is continued as long as there are constraints, which can be mapped to one variable. The propagating of the degrees of freedom is continued as long as there are variables, which have more degrees of freedom as there are still non mapped constraints in the adjacent constraint-set. For the remaining not yet assigned constraint set an assigning is made by the maximal matching method.

Propagating of Constraints

Algorithm p-con:

```
K := {c∈C | δnone(C )=1}
while K≠∅ do begin
  c:= extract-any(K);
  for e∈Λnone( c) do e.eo := v_in;
  for v∈Γout( c) and sdf(v)=0 do begin
    for all e∈Λnone( v) do e.eo:=c_in;
    for all c∈Γout( v) and δnone( c)=1 do K:=K∪{c};
  end;
end;
```

K starts with the set of all constraints of the type fix. With this K the computation begins. The edges are directed to the adjacent variable v.

If the v has no more structured degree of freedom, all other non-oriented edges in v are directed as outgoing. If a constraint c adjacent to v has now only one non-oriented edge, c is included in K.

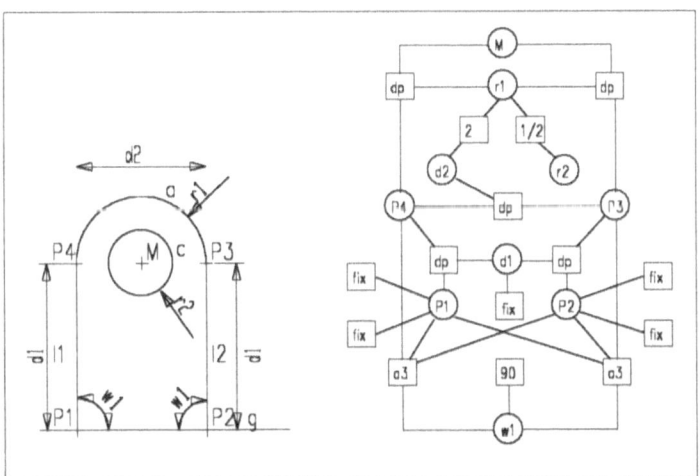

Figure 7

To show the method, we give an example. **Figure 7** shows the model with the undirected graph. The lines l1 and l2 have the same length and are perpendicular to g. The radius r2 of c is half of the radius r1 of c. The arc a is tangential to the lines l1 and l2.

The algorithm starts with K built from the constraints of the types fix and 90-degree, which is a special type of fix. Figure 8 shows the propagating. After applying the constant-constraints of P1 and P2, the angles w1 and d1 no longer have a structured degree of freedom, so the remaining edges can be oriented as c_in-edges to the adjacent dp- and a3-constraints, whose constraint-nodes are then included in the set K (left). The propagating of these constraints allows it to include the dp-constraint between P3 and P4 in K (middle). Its propagating allows it to propagate the constraints holding the radii-relations. At the end the constraints adjacent to M can be propagated (right).

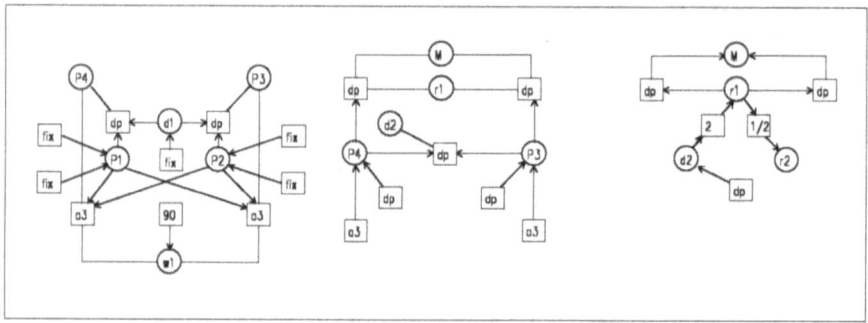

Figure 8

After applying p-con there are no constraints, to which variables can be applied in a deterministic manner. The remaining sub-graph is invariant to the order in which the constraints were chosen of K.

Propagating the degree of freedom

```
algorithm p-sfd
K := {v∈V | 0<δ_none(v )<=sdf(v)}
while K≠∅ do begin
  c:= extract-any(K);
  for all e∈Λ_none( v) do e.eo := v_in;
  for all c∈Γ_in( v) do begin
    for all e∈Λ_none( c) do e.eo:=c_in;
    for all w∈Γ_in( c) and ϑ_none( w)<=sdf(w) do K:=K∪{w};
  end;
end;
```

K is the set of variables for which edges can still be directed. The edges from the adjacent constraints are oriented as v_in. The remaining edges of the adjacent constraint can now be oriented as c_in. The variables which are adjacent over this c_in edges are included in the set K, if their degree of freedom can now be propagated. After applying the p-sdf algorithm we get a valid constraint-graph which has no sdf, which ca be propagated. The resulting graph is independent of the order in which the nodes are extracted from K.

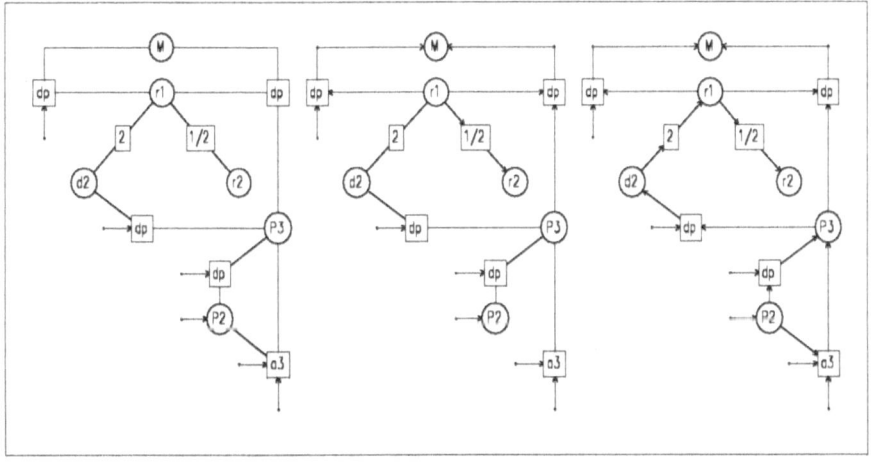

Figure 9

To illustrate the method, we continue the above example. The left part of figure 9 shows the rest of the non-oriented graph after propagating the constant constraints w1, P1 and P4. The arrows represent the oriented edges, which are not shown. At the start K={r2,M}, because only r2 and M meet the requirements for propagation. The middle part shows the constraint graph after propagating r2 and M. The adjacent constraints are assigned and the edges oriented. The variable r1 is included in K. After propagating r1, d2 meets the requirement and then P3.The right part shows the complete oriented constraint-graph after propagating P3.

Determining the maximal constraint matching

The algorithm to determine the matching is adapted from the Edmonds method for bipartite graphs [MH 90].

algorithm match

```
R := {c∈C | δnone(c )≠0}
for all c∈R do
for all e∈Λnone( c) do e.eo:=c_in;
while R≠∅ do begin
   c:= extract-any(R);
   if find/c,v) then apply(c,v);
end;
```

In the first step all incident non-oriented edges are oriented as in-coming edges. By this the graph becomes completely oriented. To find a path in the graph, which extends the matching find(c,v) looks for a variable, which still has an sdf. Then apply(c,v) inverts the orientation on the path from c to v.

After these three steps we get a graph which is completely oriented, and the variable and constraint nodes have a valid orientation. This graph can be propagated in topological order, if there are no cycles. If there are cycles, the variables on the cycles must be solved simultane-ously by solving an equation system either algebraically or with itera-tion. The propagation outside the cycles is possible, because the con-straints can always be computed with any edge as outgoing edge.

Conclusions

Two systems for constraint geometric design were explained. The first system differs from other systems mainly by the contained segments and the introduction of segments with alternatives and with repeti-tions. The second system easily allows changing the input and output variables of a geometric system.

References

[BeRo93] Berling R., Rosendahl M., 'Geometry Modelling using Di-mensional Constraints'. In CARs&FOF'93, 9th International Conference on CAD/CAM, Robotics & Factories of the Future, Newark, NewJersey, USA, August 18-20,1993

[Ber96] Berling R. 'Eine Constraint-basierte Modellierung für Geome-trische Objekte', Dissertation,1996.

[DRB93] Du C, Rosendahl M. and Berling R., 'Variation of Geometry and Parametric Design', Proc. 3rd. International Conference on CAD and Computer Graphics, Beijing, Aug. 23-26, 1993, pp 400-405, International Academic Publishers, 1993

[MH90] McHugh J., 'Algorithmic Graph Theory', Prentice-Hall, New Jersey 1990.

[RBD96] Rosendahl M., Berling R.,Du C., 'A Generalised Segment Con-cept',Proceedings of the Dagstuhl Seminar 'CAD Tools for Products' August 95,Springer.

[RSV89] Roller D., Schonek F. and Verroust A., 'Dimension-driven Geometry in CAD: a survey' in Strasser W. and Seidel H.-P. (Eds.) Theory and Practice of Geometric Modelling, Springer Verlag, 1989, pp 509-523.

[SeGo87] Serrano D. and Gossard D., 'Constraint Management in Conceptual Design' in Sriram D. and Adey R.A. (Eds.) Knowledge Based Expert Systems in Engineering: Planning and Design. Computational Mechanics Publications, 1987.

[VSR92] Veroust,A. Schonek,F., Roller,D., 'Rule oriented Method for Parametrized Computer-Aided Designs, CAD, Vol.24, No.10, pp531-540.

Geometric Constraint Decomposition

Christoph M. Hoffmann*
Andrew Lomonosov†
Meera Sitharam†

We present a flow-based method for decomposing the graph of a geometric constraint problem. The method fully generalizes degree-of-freedom calculations, prior approaches based on matching specific subgraph patterns, as well as prior flow-based approaches. Moreover, the method generically iterates to obtain a decomposition of the underlying algebraic system into small subsystems.

Introduction

Informally, a geometric constraint problem consists of a (finite) set of geometric elements and a (finite) set of constraints between them. The geometric elements are drawn from a fixed universe such as points, lines, circles and conics in the plane, or points, lines, planes, cylinders and spheres in 3-space. The constraints are logical constraints such as incidence, tangency, perpendicularity, etc., or metric constraints such as distance or angle.

The solution of a geometric constraint problem is an instantiation of the geometric elements such that all constraints are satisfied. Here, it is understood that such a solution is in a particular geometry, for example the Euclidean plane, the sphere, or Euclidean 3-space.

---

*Department of Computer Science, Purdue University, West Lafayette, Indiana 47907-1398. Supported in part by NSF Grants CDA 92-23502 and CCR 95-05745, and by ONR Contract N00014-96-1-0635.

†Department of Computer Science, Kent State University, Kent, Ohio 44242. Supported in part by NSF Grant CCR 94-09809.

In general, every geometric constraint problem can be translated mechanically into a set of nonlinear equations. The equations are usually algebraic, and nonalgebraic formulations involving trigonometric functions can be avoided in nearly all cases. The equations express the constraints, the variables are the coordinates of the geometric elements, in a suitable coordinate system.

This foundational perspective does not disclose that computing a solution of the nonlinear system is computationally challenging, and except for simple constraint systems, the problem cannot be solved in practice without further machinery. Direct approaches to algebraically processing the *entire* system include standard methods for ideal membership and locating solutions in algebraically closed fields, for example using Gröbner bases [RF96] or the Wu-Ritt method [CGZ96]; numerous algorithms and implementations for solving over the reals based on the methods of, for example, [Can93, Ren92, Col75, Laz81] etc.; and algorithms for decomposing and solving sparse systems of polynomial equations based on [JC93, Stu93, Kho78] etc. They apply to general systems of polynomial equations, hence have at least exponential time complexity, are slow in practice, and do not exploit special properties of geometric constraints. They should be used primarily for preprocessing or solving small, compact subsystems. The problems would be further compounded if we allowed constraints that must be expressed as inequalities, such as "point P is to the left of the oriented line L in the plane." Such additions necessitate using cylindrical algebraic decomposition based techniques [Col75], such as in [DG88, Laz91, Wan93] which have a theoretical worst-case complexity of $O(2^{n^2})$, where n is the algebraic size of the problem, or nonlinear optimization techniques, all of which are slow enough in practice that they do not represent a viable option for large problem sizes.

Problem Decomposition

If we wish to solve a geometric constraint problem, without discarding the algebraic perspective outright, we will need an effective way in which to decompose the equation system. Such decomposition could be done purely algebraically, using symbolic computation techniques such as those for sparse systems discussed in the previous section, but doing so is inferior, in many cases, to a generalized degree-of-freedom analysis. The latter approach first translates the constraint problem into a graph in which the graph vertices are the geometric elements and the constraints between them are incident edges.

In the constraint graph, it is possible to abstract all geometric elements by numbers that characterize the number of generalized coordinates that must be determined in order to instantiate the element, i.e, the degrees of freedom of the geometric element. In planar Euclidean geometry, for example, a point and a line each have two degrees of freedom, a circle has three, an ellipse five, and so on. These numbers are affixed to the graph vertices as *weights*. Similarly, the constraints can be abstracted by the number of coordinates that are fixed by the constraint, usually the number of equations that express the constraint algebraically. Again, in the Euclidean plane, an incidence constraint between a point and a line determines one coordinate value, whereas an incidence constraint between two points determines two. Therefore, the respective number is affixed to the graph edge representing the constraint, as *weight* of the edge.

As was shown in [HLS97], a key step in the decomposition of the constraint graph into generically solvable subsystems is to find minimal dense subgraphs. A *weighted undirected graph* is a graph where every vertex and every edge has a nonzero integer weight.

Let G be a weighted undirected graph, $G = (V, E, w)$, with n vertices V and m edges E, where $w(v)$ are the weights of vertices and $w(e)$ are the weights of edges. We want to find an vertex-induced subgraph $A \subseteq G$ such that

$$\sum_{e \in A} w(e) - \sum_{v \in A} w(v) > K \qquad (1)$$

Such a subgraph A is called *dense*.

More precisely, we want to find a minimal dense subgraph, that is, a dense subgraph A such that A does not contain a proper dense subgraph B. The related problem of finding the **minimum** dense subgraph A is shown to be NP-hard.

Broadly speaking, a minimal dense subgraph corresponds to a subproblem of the geometric constraint problem that can be solved separately in the generic case, and therefore can be used to decompose the nonlinear system of equations. More precisely, the remaining degrees of freedom in the system will be $-(K+1)$. Thus, if $K = -1$, then the constraint system is rigid with respect to a global coordinate system and is therefore fully constrained. For a constraint problem in the plane, we often deal with $K = -4$ where the geometric entities are rigid with respect to each other but can be translated and rotated in the plane with respect to a global coordinate system. Such a constraint problem is well-constrained.

As soon as a subgraph of the appropriate density has been found, the corresponding geometric objects can be placed rigidly with respect to each other (or with respect to a global coordinate system) using only the constraints between them. The solver would then continue by condensing the constraint graph, coalescing the placed elements into a new graph vertex and suitably inducing edges to the other vertices. The description of this process is given in Section 5.

Prior Work on Constraint Graph Analysis

Prior attempts at a degree-of-freedom analysis for constraint graphs often concentrated on recognizing specific dense subgraphs of known shape, such as the triangles of [BFH+95, Owe91, Owe93] or the patterns of [BFH+95, HV94, HV95]. This approach has limited scope: certain constraint problems can be decomposed very efficiently, but many well-constrained problems cannot be decomposed and the solvers give up on them. The scope can always be extended by increasing the repertoire of patterns of dense subgraphs. However, doing so results in greater combinatorial complexity and eventually makes an efficient implementation too difficult.

More general attempts reduce the recognition of dense subgraphs in a degree-of-freedom analysis to a maximum weighted matching problem in bipartite graphs using methods from, e.g., [LM96]. A variation [AJM93] of this approach does not use a degree-of-freedom analysis and directly deals with algebraic constraints. In this case, a maximum cardinality bipartite matching is used, since no weights are required. The approach then relies on decomposing into strongly connected components, and one can attempt to generalize it to a weighted version required for a general degree-of-freedom analysis either by replicating vertices, or by retaining weighted vertices, as in [Pab93]. We discuss later why all of these approaches are incomplete and less efficient than the approach presented here. In particular, having found the required matching, or maximum flow, finding a dense subgraph requires significant additional work; it becomes difficult to isolate minimal dense subgraphs, and the approaches only work for density 0, i.e $K = -1$, and do not generalize to arbitrary densities or values of K. The general approach of [Hsu96] appears to be exponential.

A different approach to constraint graph analysis uses rigidity theorems; e.g., [CH88, Hav91]. Corresponding decomposition steps may be nondeterministic or require difficult symbolic computations when computing a solution.

Dense Subgraphs

In decomposing a geometric constraint system, one would ideally like to locate the smallest possible subsystems. This corresponds generically to finding a minimum dense subgraph of the weighted constraint graph. A provable bound on the size of the minimum dense subgraph, for example, would permit a feasible search for one. Unfortunately (even with the common assumption that the weights of the vertices and edges are bounded) it is easy to construct graphs that have arbitrarily large minimum dense subgraphs. In fact, the next section renders a feasible search unlikely, by showing that the problem (when the weights of the vertices and edges is unbounded) is NP hard.

Finding a Minimum Dense Subgraph is NP-hard

For a weighted undirected graph G and a constant p, let *SMALL-DENSE* be the problem of deciding whether there is a dense subgraph (without loss, for density zero) in G with at most p vertices.

For an input undirected graph G, let *CLIQUE* be the problem of deciding whether there is a complete subgraph in G with at least p vertices. Recall that *CLIQUE* is NP-complete. A polynomial time reduction of *CLIQUE* to *SMALL-DENSE* shows that *SMALL-DENSE* is NP-hard:

Proposition

SMALL-DENSE is NP-complete.

Proof

That *SMALL-DENSE* is in NP is obvious. Let $G' = (V, E, w)$ be a weighted undirected graph. Every vertex has the weight $w(v) = p(p - 1)/2$, and every edge has the weight $w(e) = p$. Let $G = (V, E)$ be the corresponding undirected graph, and (G, p) an instance of *CLIQUE*. If G has a clique of size p, then the corresponding subgraph of G' is dense. Conversely, let S be a dense subgraph of G' of size s at most p. Because of the weights, $s < p$ is not possible. Therefore, since $s = p$, there must be $p(p - 1)/2$ edges in S; that is, there is a clique of size p in G. $\square$

A Greedy Algorithm

Next we give the straightforward greedy algorithm for finding dense (not necessarily minimum, or even minimal), subgraphs and show that it runs for exponential time in the worst case.

Definition

For $A \subseteq G$ define the *density* function $d(A)$ as

$$d(A) = \sum_{e \in A} w(e) - \sum_{v \in A} w(v)$$

The idea of the greedy algorithm is to start with an empty list of sub-graphs L. Subgraphs $A \in L$ are built by adding one vertex at a time and computing the density function $d(A \cup \{v\})$. One could expect that L is simply the list of all possible subgraphs considered so far and that its size would increase exponentially. However, it turns out that it is sometimes possible to ignore $A \cup v$ or A, so that the size of L does not always increase when adding a vertex.

Let B be the set of all vertices that were already considered. If $A \cup \{v\}$ is dense then we are done. If not, then if $d(A \cup \{v\}) \geq d(A)$ we can replace A by $A \cup \{v\}$, thus keeping the size of L unchanged. If

$$d(A \cup \{v\}) < d(A)$$

and

$$d(A \cup \{v\}) + \sum_{e=(u,u'),u\in B,u'\notin B} w(e) \leq d(A)$$

then we will not add $(A \cup \{v\})$ to L and keep A in L, so the size of L is unchanged. Therefore, we only increase the size of L by adding $A \cup \{v\}$ when $d(A \cup \{v\}) < d(A)$ and $d(A \cup \{v\}) + cut(v, G - B) > d(A)$

Pseudocode for the Greedy Algorithm

```
1.   L = {}
2.   B = {}
3.   for every v in G
4.       B = B ∪ {v}
5.       for A ∈ L
6.           if d(A ∪ {v}) ≥ 0 then
7.               return A ∪ {v}
8.           else if d(A ∪ {v}) ≥ d(A) then
9.               replace A by A ∪ {v}
10.          else if d(A ∪ {v})+ cut(v, G − B) > d(A) then
11.              L = L ∪ (A ∪ {v})
12.          endif
13.      endfor
14.  endfor
```

Performance of the Greedy Algorithm

When G is not dense, the algorithm may require exponential time. For
example, if G is the rectangular grid, the weight of all vertices is 2, and
the weight of all the edges is 1, then the algorithm creates a list L of
exponential length.

Prior Work Using Flow-Based Approaches

Suppose that we want to find a *most* dense subgraph $A \subseteq G$, i.e, one for
which $d(A)$ is maximum. We could maximize, over subgraphs A of G, the
expression

$$d(A) + \sum_{v \in G} w(v) = \sum_{e \in A} w(e) + \sum_{v \notin A} w(v) \tag{2}$$

or, equivalently, minimize

$$\min_{A \subseteq G}(\sum_{e \notin A} w(e) + \sum_{v \in A} w(v)) \tag{3}$$

To do this, consider a bipartite graph $\tilde{G} = (M, N, \tilde{E}, w)$ associated with
the given graph $G = (V, E, w)$. The vertices in N are the vertices in V
and the vertices in M are the edges in E. Moreover, the edges of $\tilde{G}$ are
$\tilde{E} = \{(e, u), (e, v) \mid e = (u, v), e \in E\}$. The weights w now appear on the
vertices of $\tilde{G}$. Maximizing the expression (2) reduces to finding a maximum weighted independent set in the bipartite graph $\tilde{G}$, or, equivalently,
the minimum weight vertex cover.

There are two ways to try to find the minimum weight vertex cover.
The minimum *cardinality* vertex cover in a bipartite graph can be identified with a maximum cardinality matching and can be found using network flow in $O(\sqrt{n}m)$ time [ET75]. To take advantage of this algorithm,
however, we need to replicate edges and vertices by the corresponding
weights. For example, a vertex of weight 3 is tripled. In this larger graph,
we find a minimum cardinality vertex cover, and then we try to locate a
corresponding minimum weight vertex cover in $\tilde{G}$ and the corresponding
dense subgraph in the graph G. No efficient method is known for the
latter part.

The unweighted version of bipartite matching in $\tilde{G}$ can be used naturally when variables are directly represented as vertices in G; the algebraic equations are assumed to kill only one degree of freedom, and
are represented as edges in G (instead of analyzing general degrees-of-
freedom). This again results in a constant factor increase in the size of
the graph. The matching naturally induces a direction on the edges of

the original graph G, and the strongly connected components provably yield a decomposition into minimal dense subgraphs in the case when the density is chosen to be exactly 0 ($K = -1$). This approach was used in [AJM93], and in fact, gives a natural way of decomposing the entire graph into minimal dense subgraphs. However, it is not clear how to extend the algorithm for general K, or for general, weighted, degree of freedom analysis, with constraints that kill more than one degree of freedom.

An attempt to directly extend this method to general degree of freedom analysis, i.e, to weighted graphs can be found in [Pab93], although it preceded [AJM93]. The method is also a flow-based method and is superficially similar to ours, however, it differs in crucial aspects. Create a source and sink in $\tilde{G}$ corresponding to the M and the N sets of vertices respectively, assigning capacities to M and N corresponding to the weights of the edges and vertices in G respectively. Now a *maximum* flow is found, which corresponds to a "generalized" matching and induces a natural direction on the edges of the graph G as in [AJM93]. Unlike in the unweighted case, however, the strongly connected components are neither guaranteed to correspond to minimal dense, or even dense subgraphs of G, nor do they provide a natural decomposition method, *even for the zero-density case*, i.e, $K = -1$.

A second method for dealing with weights is to search for a minimum weighted vertex cover in $\tilde{G}$ by solving the maximum (vertex) weighted bipartite matching problem. A maximum (edge) weighted bipartite matching problem can be solved in $O(\sqrt{n}m \log n)$ time for bounded weights, [RAO93]. This trivially gives a solution to the maximum (vertex) weighted bipartite matching problem. The catch is that, unlike in the unweighted case, a minimum weighted vertex cover does *not* correspond directly to a maximum weighted matching. Having found a maximum weighted matching, a significant amount of work is needed to obtain the minimum weight vertex cover, and, from it, the corresponding dense subgraph in G.

Summarizing, the above approaches have the following disadvantages.

1. The maximum (weighted) matching or maximum flow in $\tilde{G}$ does not directly yield dense subgraphs in G.

2. We need only *some* subgraph of a specific density, not necessarily a *most* dense one. Hence, in the flow based approaches, it is *not* necessary to find the maximum flow.

3. The above approaches provide no natural way of finding a minimal dense subgraph for arbitrary weighted graphs, for arbitrary densities or values of K.

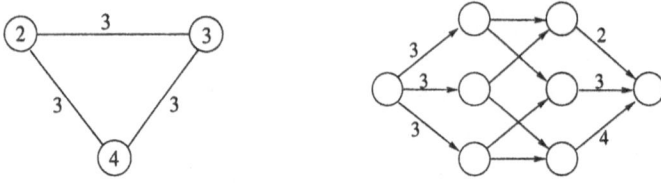

Figure 1: Constraint graph (left) and associated network (right).

We develop a more efficient method analogously based on a different optimization problem (see [Law76]), but which will be seen to address all of these drawbacks.

Finding a Dense Subgraph

We devise a flow-based algorithm for finding subgraphs that have density 1; i.e., $K = 0$ in Equation (1). We discuss the case of other densities the following section.

Construction of the Network

From the graph G, construct a bipartite *directed* network $G^* = (M, N, s, t, E^*, w)$, where M, N and E^* are as in $\tilde{G}$. The source s is connected by a directed edge to every node in M, and every node in N is connected by a directed edge to the sink t. The capacity of the network edge (s, e), $e \in M$, is the weight $w(e)$ of the edge e in G. The capacity of the network edge (v, t), $v \in N$, is equal to the weight $w(v)$ of the vertex v in G. The capacity of the network edge (e, v), $e \in M$, $v \in N$, is infinite. There are no other network edges. See also Figure 1.

Notice that the construction of the network extends to hypergraphs representing ternary or other constraints, where each hyperedge involves an arbitrary number of vertices.

A minimum cut in G^* *directly* defines a subgraph A that minimizes Expression (3). It can be found as the max flow using a netflow algorithm. Now we are only interested in finding a dense subgraph and not necessarily the *most* dense one. So, we are interested in a small enough cut in G^*, not necessarily the smallest one. Thus, to find a dense subgraph, there should be an algorithm that is faster than a general maximum flow

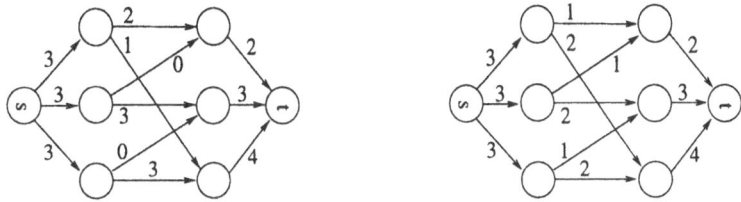

Figure 2: Two different flows for the constraint graph of Figure 1

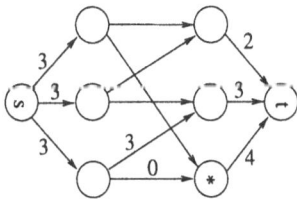

Figure 3: Initial flow assignment that requires redistribution later

(or minimum cut) algorithm.

The algorithm given in the next section relies on a subtle, but crucial modification of the incremental max flow algorithm which seems tailor-made for the current application in that it simultaneously addresses all the drawbacks of the previous algorithms mentioned in the previous section.

The Dense Algorithm

The idea of the algorithm (Algorithm Dense below) is to start with the empty subgraph G' of G and add to it one vertex at time. When a vertex v is added, consider the adjacent edges e incident to G'. For each e, (for ease of exposition, we assume the edges are binary) try to distribute the weight $w(e)$ to one or both of its endpoints without exceeding their weights; see also Figure 2. As illustrated by Figure 3, we may need to redistribute some of the flow later.

If we are able to distribute all edges, then G' is not dense. If no dense subgraph exists, then the algorithm will terminate in $O(n(m+n))$ steps and announce this fact. If there is a dense subgraph, then there is an edge whose weight cannot be distributed even with redistribution. The last vertex added when this happens can be shown to be in all dense subgraphs $A \subseteq G'$. Distributing an edge e in G now corresponds to pushing a flow equal to the capacity of (s, e) from s to t in G^*. This is possible either directly by a path of the form $\langle s, e, v, t \rangle$ in G^*, or it might require flow redistribution achieved by a standard search for augmenting paths [FF62], using network flow techniques, see Figures 4, 5. Note that the search for augmenting paths takes advantage of the fact that the flow through each vertex in M is distributed to exactly 2 vertices in N (lines 4-7) in Algorithm Distribute. While this decreases running time by a constant factor, it doesn't affect complexity.

If there is an augmenting path, then the resulting flows in G^* provide a distribution of the weight of each edge e in the current subgraph G' consisting of the examined vertices and edges of the original graph G as follows: the weight $w(e)$ of each edge e connecting the vertices a and b is split into two parts f_e^a and f_e^b such that $f_e^a + f_e^b = w(e)$ and, for each vertex $v \in G'$, $\sum_{e=(v,*)} f_e^v \leq w(v)$.

If there is no augmenting path for the residual flow on (s, e), i.e, the flow $w(e)$ is undistributable, then a dense subgraph has been found and is identified based on the flows in G^* starting from e.

Algorithm Dense

1. $G' = \emptyset$.
2. **for every** vertex v **do**
3.   **for every** edge e incident to v and to G' **do**
4.   <u>Distribute</u> the weight $w(e)$ of e
5.   **if** not able to distribute all of $w(e)$ **then**
6. A = set of vertices labeled during <u>Distribute</u>
7.   **goto** Step 12
8.   **endif**
9.   **endfor**
10.   add vertex v to G'
11. **endfor**
12. **if** $A = \emptyset$ **then** no dense subgraph exists
13. **else** A is a dense subgraph

Algorithm Distribute searches for augmenting paths in G^* to reach the required flow and the labeling. It repeats a Breadth First Search for augmentation until all of $w(e)$ has been distributed or until there is no augmenting path. The technique is similar to the one used in the max-flow algorithm in [Law76].

Algorithm Distribute

Input: $(G^*, f, edge)$, where $G^* = (N, M, s, t, E^*, w)$, f is a set of
flows f_e^v and *edge* is the edge that is being distributed.

0. Initialize $scan(v) = 0, label(v) = 0, scan(e) = 0$
    $label(e) = 0$ for all $v \in N, e \in M$
1. $vert = 0$, $capvert = 0$
2. $label(edge) = 1$, $pathcap(edge) = w(edge)$
3. **while** $(w(edge) > \sum_v f_{edge}^v)$ **or** not all labeled nodes
                                        have been scanned
4.   **for** all labeled $e \in M$, with $scan(e) = 0$
5.   label unlabeled neighbors of e (i.e $v \in N$)
6. $scan(e) = 1$, $pred(v) = e$, $pathcap(v) = pathcap(e)$
7.   **endfor**
8.   **for all** labeled $v \in N$ with $scan(v) = 0$
9.   **if** $\min(w(v) - \sum_e f_e^v, pathcap(v)) > capvert$ **then**
10. $vert = v$, $capvert = min(w - \sum_e f_e^v, pathcap(v))$
11.   **else**
12.   label all unlabeled $e' \in M$ s.t $f_{e'}^v > 0$
13.   **endif**
14. $scan(v) = 1$
15.   **endfor**
16.   **if** $vert > 0$ **then**
17.   An augmenting path from s to t has been found:
            backtrack from *vert* using $pred()$ and change the
            values of f_e^v as required.
18.   **for all** $e \in M, v \in N$
19. $label(e) = 0, scan(e) = 0, label(v) = 0, scan(v) = 0$
20.   **endfor**
21. $vert = 0, capvert = 0, label(edge) = 1$
22. $pathcap(edge) = w(edge) - \sum_v f_{edge}^v$
23.   **endif**
24. **endwhile**

Lemma 1
Let G^* be the bipartite network constructed from G, and $e \in M$. If, after
checking all possible augmenting paths originating at e, the flow through
(s, e) is less than the capacity of (s, e), and $A = (E_A, V_A)$ is the set of
edges and vertices labeled after the search for an augmenting path, then
$d(A) > 0$.

Proof: A is a subgraph of G because for every labeled edge $e \in E$ both
of its vertices will be labeled. For all $v \in V_A$, the network edges (v, t)

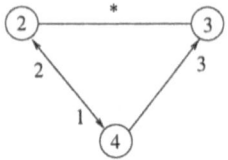 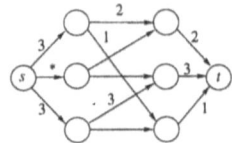

Figure 4: Current graph G' and corresponding network G^*, the edge marked by asterisk is currently being distributed

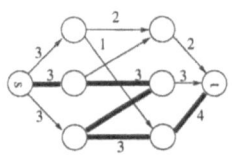 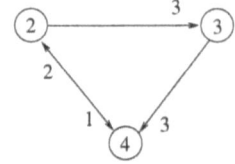

Figure 5: The augmenting path and the distribution of edges in original graph G'

are saturated, otherwise there is an augmenting path from e to v and the flow through (s, e) can be increased. Let f be the maximum flow through (E_A, V_A). Since all (v, t) are saturated, $f = \sum_{v \in A} w(v)$, but since at least one edge (s, e) is not distributed $f < \sum_{e \in A} w(e)$; therefore $d(A) = \sum_{e \in A} w(e) - \sum_{v \in A} w(v) > 0$. $\square$

The correctness of Algorithm Dense follows from the above Lemma, since if the graph contains any dense subgraph, Algorithm Dense will find it.

Complexity Analysis

In the worst case, constructing an augmenting path labels at most $m + n$ nodes. Since the algorithm stops when the total edge weight exceeds the total vertex weight, the total edge weight that is distributed is at most the total vertex weight times a constant bound b. Each augmentation increases the flow by least 1 unit. Therefore, the number of augmentations cannot exceed $O(n)$. Hence, Algorithm Dense has complexity $O(n(m+n))$.

Finding a Minimal Dense Subgraph

Let $G' = (V', E')$ be the subgraph already examined by Algorithm Dense. That is, assume that the vertices V' have been examined and the weight

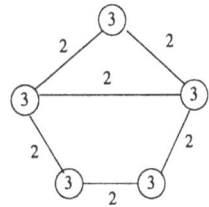

Figure 6: This subgraph is dense for $K = -4$, so is upper triangle

$w(e)$ of all induced edges e has been distributed. Let v be the first vertex that is about to be examined next, such that the weight of one of its incident edges e adjacent to G' cannot be distributed. Let $V_A \subseteq V'$ be the set of vertices labeled while trying to distribute $w(e)$, (which includes the vertex v), and let A be the subgraph induced by V_A. By Lemma 1, A is a dense subgraph.

Lemma 2

Every dense subgraph of A contains v.

Proof: Let A' be a dense subgraph of A not containing v. Then there should be an edge $e \in A'$ such that e was not distributed before v was considered. However, this contradicts our assumption that all edges in G' have been distributed. $\square$

Remark

Similarly, if $(v, v_1), (v, v_2), ..., (v, v_k)$ are undistributed edges of v then every dense subgraph of A contains at least one edge from this list. If $k = 1$ then every dense subgraph of A contains (v, v_1).

Proposition 3

If the amount of undistributable flow, i.e, the density of A is $d(A)$ and A' is a dense proper subgraph of A, then $0 < d(A') < d(A)$ (in general, $K < d(A') < d(A)$).

Proof: Note that the excess flow comes from the edges incident to v. Suppose $A' \subseteq A$ is dense and $d(A') \geq d(A)$. By Lemma 2, A' contains v. Consider the relative complement A^* of A' with respect to A. Then $d(A^*) \leq 0$, which implies that the vertices of A^* could not have been labeled after distributing the flow of the edges of v. Since all vertices in V_A are labeled, we know that $A = A'$. $\square$

Corollary 4

If $d(A) = 1$ then A is minimal. In general, if $d(A) = K + 1$, then A is minimal.

In particular, when $K = 0$, well-constrained or underconstrained problems have $d(A) \leq 1$. Then, by Corollary 4, we know that the subgraph found by Algorithm Dense is minimal. Moreover, if overconstrained problems are rejected, then a first test for overconstrained would be to determine $\sum_{e \in G} w(e) - \sum_{v \in G} w(v) > 1$ in linear time. This test would reject many overconstrained problems. The remaining cases would be found by noting whether $d(A) > 1$ when Algorithm Dense terminates.

We may accept consistently overconstrained problems. In that case, the graph A may have to be analyzed further to extract a minimal dense subgraph. We now develop a method for performing this extraction, once a dense subgraph A has been found by Algorithm Dense and $d(A) > 1$. The algorithm to be developed post-processes only the subgraph A.

Without loss of generality, assume that A contains the vertices $\{v_1, \ldots, v_l, v_{l+1}\}$, and v_{l+1} was the last vertex examined when A was found. The density $d(A)$ is the total undistributed weight of the edges between v_{l+1} and $\{v_1, \ldots, v_l\}$. We begin with the knowledge of a subgraph B of A that is contained in *every* dense subgraph of A. By Lemma 2, B contains initially the vertex v_{l+1}. The algorithm to be developed is to determine either an enlargement of the graph B, or else a reduction of the graph A.

We perform the following step iteratively. Choose a vertex $v_k \notin B$ from A. Determine the quantity $c = d(A) - w(e') + f_{e'}^{v_k} + f_{e'}^{v_{l+1}}$ where e' is the edge (v_k, v_{l+1}). That is, c is the undistributed weight of edges in A without v_k. Remove the vertex v_k from A along with its edges. This would create unutilized capacity in the set of vertices adjacent to v_k (that are in A) through the set E_k of incident edges. The excess vertex capacity is

$$\sum_{e \in E_k} w(e) - w(v_k) - w(e') + f_{e'}^{v_k} + f_{e'}^{v_{l+1}}$$

where e' is the edge between v_k and v_{l+1}. This quantity is the total flow on the edges of v_k, distributed away from v_k. We now attempt to distribute the previously undistributed weight of the edges between v_{l+1} and $\{v_1, \ldots, v_l\} - \{v_k\}$, using redistribution if necessary. We use Algorithm Distribute on the modified network, setting the capacity of (v_k, t) to zero. There are two outcomes possible:

1. If we distribute all of c successfully into the newly created holes, or excess capacity on the vertices adjacent to v_k, then no subgraph of $A - v_k$ is dense, so v_k belongs in every dense subgraph of A, and hence gets restored into A and, moreover, gets added to B.

2. If we were unable to distribute c, then by Lemma 1, we have found a smaller dense subgraph of $A - v_k$. This new subgraph consists

only of the vertices labeled by the Algorithm Distribute in the process of distributing one of the undistributed edges adjacent to v_{l+1}. This outcome reduces the size of A. Note that, by Proposition 3, the density (and size) of the new graph A must drop by at least 1.

We repeat this process for the remaining vertices in $A - B$. We stop either when $d(B) > 0$, because then B is minimal dense, or when $d(A) = 1$, because then A is minimal dense.

Complexity Analysis

The complexity of each iteration described above is $O(n(m + n))$, since c represents the undistributed weight on at most n edges adjacent to v_{l+1} are distributed at each iteration. We can assume that the sum of capacities of the edges is constant, thus the determination of a minimal dense subgraph takes $O(n^2(m + n))$ steps. Note, however, that by Proposition 3 the actual complexity rarely reaches this upper bound.

The complexity of the iteration is reduced to $O(m + n)$ if the constraint graph has bounded valence or if $d(A)$ has an a-priori constant bound. The latter situation means that the constraint problem has a bound on the "overconstrainedness" of subgraphs, a natural assumption if the constraint problem is specified interactively and we keep track of the density of the full constraint graph. In those cases, the complexity reduces to $O(n(m + n))$ steps.

Algorithm Minimal

Comment: The input is the output of Dense, a dense (sub)graph A of G, and the distribution of edge weights f_e^a and f_e^b for each edge $e = (a, b)$. Note that v_{l+1} is the last vertex added that caused A to be found, and e' is the edge between v_k and v_{l+1}.

1. $B = \{v_{l+1}\}$
2.   **while** $d(B) \leq 0$ and $d(A) > 1$ **do**
3.   choose $v_k \in A - B$
4. $c = d(A) - w(e') + f_{e'}^{v_k} + f_{e'}^{v_{l+1}}$
5.   **for all** $v \in N$ (Removing $\{v_k\}$ from A)
6.   Let $e = (v, v_k)$
7.   remove e from M
8.   **endfor**
9.   remove v_k from N
10.   Distribute (in A) excess c from the edges of v
11.   **if** there are some undistributed edges left **then**
12.   set A = new labeled graph

13.   **else**
14.   set $A = A \cup \{v_k\}$ (as well as restoring edges of v_k)
15.   set $B = B \cup \{v_k\}$
16.   **endif**
17.   **endwhile**
18.   **output** B, **if** $d(B) > 0$, **else output** A.

The Case of $K \neq 0$

Recall that K is chosen so that the desired degrees of freedom for a well-constrained subsystem is $-(K + 1)$. Thus in the context of geometric constraint solving, $K = -1$ represents the case where the dense subgraph corresponds to elements that can be placed rigidly with respect to each other and to a global coordinate system. The case of $K > -1$ means that the subgraph is overconstrained, usually by $K + 1$ constraints, and the case of $K < -1$ means that the resulting geometric configuration has residual motion (remaining degrees of freedom) with respect to a global coordinate system.

The most important cases are when $K = -4$ in the planar case or $K = -7$ in the spatial case, signaling that the resulting geometric configuration can move as a rigid body with respect to the coordinate system. In [Fud95], this property has been exploited recursively for the purpose of cluster combination. The case $K < -1$ is also the reason why symbolic algebraic decomposition methods fail to succeed as they implicitly assume $K = -1$.

As presented so far, our algorithms satisfy Inequality (1) for $K = 0$: keep adding vertices until we are unable to distribute the edge weight/capacity. The first undistributable edge signals a dense graph A, for $K = 0$, and $d(A) > d(A - v)$, where v was the last vertex examined. We now explain how to modify the algorithm to accommodate different values of K.

The modification for $K > 0$ is trivial. Instead of exiting Algorithm Dense when an edge cannot be distributed, exit when the total undistributable edge capacity exceeds K. The computation of the total undistributable edge capacity so far is based only on the weights of the labeled edges and vertices, thus ensuring that the resulting dense graph is connected. An analogous change is in order for Algorithm Minimal. Clearly this modification does not affect the performance complexity of the algorithms.

When $K < 0$ the algorithms can also be modified without increasing the complexity. Suppose, therefore, that $K < 0$, and consider Step 4 of Algorithm Dense. If $w(v) + K \geq 0$, simply reduce the capacity of the network edge (v, t) to $w(v) + K$ and distribute $w(e)$ in the modified network. If the edge cannot be distributed, then the subgraph found in Step 6 has density exceeding K. If every incident edge can be distributed, then restore the capacity of the network edge when adding v to G'.

If the weight $w(v)$ of the added vertex v is too small, that is, if $w(v) + K < 0$, then a more complex modification is needed. We set the capacity of (v, t) to zero. Let e be an new edge to be distributed, and do the following.

1. Distribute the edge weight $w(e)$ in the modified network.
2. **if** $w(e)$ cannot be distributed **then**
3.   we have found a dense subgraph for K; exit.
4. **else**
5.   save the existing flow for Step 10.
6.   increase the flow of e by $-(w(v) + K)$
7.   **if** the increased flow cannot be Distributed **then**
8.   we have found a dense subgraph for K; exit.
9.   **else**
10.   restore the old flow. No dense subgraph found
11.   **endif**
12. **endif**

In worst case the algorithm saves and restores the flows for every edge added, which requires $O(m)$ operations per edge. Distributing the edge flow however dominates this cost since it may require up to $O(m+n)$ operations per edge; so the modification does not adversely impact asymptotic performance.

Graph Decomposition

The flow-based degree-of-freedom analysis must be applied repeatedly to constraint graphs. Two different conceptual steps are involved. When a minimal subgraph of the appropriate density has been found, then the vertices form a *cluster* of geometric elements that can be placed rigidly with respect to each other, using only the constraints represented by the edges of the subgraph. This cluster can be extended under certain circumstances by adding more geometric elements that are determined by constraints involving the cluster. After a cluster has been so extended, it

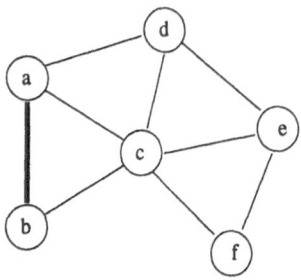

Figure 7: Constraint graph with vertices of weight 2 and edges of weight 1. The minimal dense subgraph {a,b} can be extended sequentially by the other elements, in alphabetic order.

must then be abstracted into a single geometric entity, and the rest of the constraint graph can be searched for another minimal dense subgraph.

Extended Clusters

Consider the constraint graph G of Figure 7. We assume that all vertices have weight 2 and all edges weight 1. Then the vertex set $\{a, b\}$ induces a minimal dense subgraph of G. After solving this subgraph, i.e., after assigning coordinates to the geometric elements a and b so that the constraint between them is satisfied, we can place c because its two coordinates (the weight) are determined by the two incident constraints with a and b. After so placing c, we can place, in sequence, d, e and f. Thus, all six geometric elements can be placed rigidly relative to each other.

The structure of the solution is as follows: beginning with a simultaneous system of equations, the elements of the minimal dense subgraph are placed with respect to each other (and the global coordinate system in the case of $K = -1$). In our example, two geometric elements are placed. Then, a number of geometric elements are placed, one by one, solving for each individually from as many equations as there are incident constraints on them with the previously placed elements. In our example, each element requires solving two equations in two variables.

We call a minimal dense subgraph that has been enlarged by sequential extensions an *extended dense subgraph*.

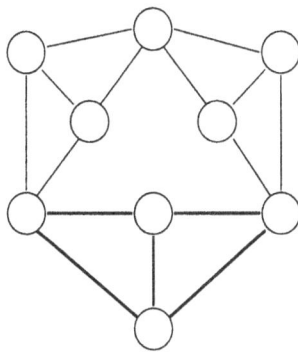

Figure 8: Constraint graph with an extended dense subgraph. Vertex weight is 2, edge weight is 1.

Recursive Application

After an extended dense subgraph has been found, the set of geometric elements that are its vertices form a rigid geometric structure. For $K = -1$, this structure is fixed with respect to the coordinate system, otherwise the structure can be moved with $-(K + 1)$ degrees of freedom.

We extend the decomposition and find other dense subgraphs, abstracting this subgraph into a geometric construct with $-(K + 1)$ degrees of freedom. This can be done applying *natural graph reduction*: Replace the dense subgraph G_0 by a vertex u of weight $-(K + 1)$ and combine all edges from a vertex in G_0 to a vertex w not in G_0. The weight of the induced edge (u, w), in the reduced graph, is the sum of weights of the edges that have been combined. After the reduction, another dense subgraph is found. Reduction ends when the graph has been reduced to a single vertex.

For example, consider the graph of Figure 8. All vertices have weight 2, all edges have weight 1. The four vertices connected by the heavy edges constitute an extended dense subgraph. After reduction of the extended dense subgraph we obtain the graph of Figure 9. Again, an extended cluster is found and indicated by the heavy edges in the graph. Reducing this cluster, we obtain the graph of Figure 10. In this graph, a minimal cluster is found as indicated, and further reduction yields the graph of Figure 10. Here, the entire graph is minimal dense, so the last reduction yields a graph that consists of a single vertex.

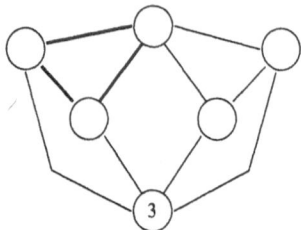

Figure 9: Constraint graph after reducing the subgraph. The cluster vertex has weight 3.

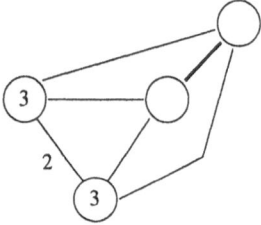

Figure 10: Constraint graph after two subgraph reductions. The cluster vertices have weight 3 each, the edge between them has weight 2.

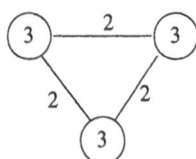

Figure 11: Constraint graph after three subgraph reductions.

Church-Rosser Property

A constraint graph G is *well-constrained* if the density of G is $d(G) = K+1$ and there is no vertex-induced subgraph of density greater than $K + 1$, where K is the parameter chosen in the inequality (1).

If a well-constrained geometric constraint graph can be reduced, by a sequence of the reduction steps described before, to a single vertex, then the geometric constraint problem is solved in the generic case by the corresponding decomposition of the nonlinear equation system. Therefore, we would like to show that:

> A well-constrained graph can be reduced iteratively to a single vertex, no matter which dense subgraph is chosen at each step of the reduction sequence by our algorithm.

This property would eliminate the need to choose a particular reduction sequence for successfully decomposing a constraint system. As we show, this property holds. Intuitively, the property follows from the fact that subgraph reduction preserves the density of the graph.

The reduction algorithm considers a graph G_i, finds a well-constrained subgraph H_{i+1} and reduces G_i to G_{i+1}. We denote such a reduction step by

$$G_i \xrightarrow{H_{i+1}} G_{i+1}$$

The reduction terminates when no dense subgraph can be found.

Theorem Let G be a well-constrained constraint graph. Consider any complete reduction sequence

$$G = G_0 \xrightarrow{H_1} G_1 \xrightarrow{H_2} \cdots \xrightarrow{H_m} G_m$$

produced by our decomposition method, where the reduction halts with G_m. The subgraphs H_i are nontrivial (of size greater than 1) and have been located by the Algorithm Dense. Then $|G_m| = 1$.

Proof. The proof follows immediately from the fact that the process of condensing does not change the density. Let H be a subgraph of G. Then

$$d(G) = d(H) + d(G - H) + \sum_{u \in H, v \in G-H} w((u,v))$$

In the reduction

$$G \xrightarrow{H} G'$$

we replace H with a vertex h whose weight is $w(h) = -d(H)$. Moreover, we remove the edges (r, u), where $r \in H$ and $u \notin H$ and add the edge weight to the edge (h, u) of G'. Thus $d(G) = d(G')$, and therefore $d(G) = d(G_0) = d(G_1) = \ldots = d(G_m)$. Since G is well-constrained, it follows that G_m is also well-constrained. Since the decomposition method halts only when the current (reduced) graph does not have a well-constrained subgraph of size greater than 1 (by the correctness of Algorithm Dense), we can assume that G_m has no well-constrained subgraph (including itself) of size greater than 1. Therefore, $|G_m| = 1$, thus proving the theorem. $\square$

Note that the theorem implies that the graph rewrite system defined by the reductions of constraint graphs has the Church-Rosser property for well-constrained graphs. For other constraint graphs the property is not implied since two graphs with the same density need not be isomorphic.

Conclusion

The flow-based approach presented here is a general and efficient approach at analyzing constraint problems and decomposing them into subproblems that can be solved in isolation. Of particular importance is that the method has the Church-Rosser property for well-constrained problems, because in general the choice of subproblem and the resulting decomposition is not unique.

We are currently incorporating the approach into a recursive graph decomposition that ends with a single vertex after several reduction steps. Of special interest here is to find suitable strategies that are responsive to the needs of the subsequent solving phase in which instance equations are formulated and solved. Even though two different graph decompositions may look equivalent, as graphs, they could lead to equation systems of different size.

References

[AJM93]   S. Ait-Aoudia, R. Jegou, and D. Michelucci. Reduction of constraint systems. In *Compugraphics*, pages 83–92, 1993.

[BFH+95] W. Bouma, I. Fudos, C. Hoffmann, J. Cai, and R. Paige. A geometric constraint solver. *Computer Aided Design*, 27:487–501, 1995.

[Can93]   J. Canny. Improved algorithms for sign determination and existential quantifier elimination. *Computer Journal*, 36(5), 1993.

[CGZ96]   S. C. Chou, X. S. Gao, and J. Z. Zhang. A method of solving geometric constraints. Technical report, Wichita State University, Dept. of Computer Sci., 1996.

[CH88]   G. Crippen and T. Havel. *Distance Geometry and Molecular Conformation*. John Wiley & Sons, 1988.

[Col75]   G.E. Collins. Quantifier elimination for real closed fields by cylindrical algebraic decomposition. In *Lect. Notes in CS, Vol 33*, pages 134–183. Springer-Verlag, 1975.

[DG88]   N. Vorobjov D. Grigoriev. Solving systems of polynomial inequalities in subexponential time. *J. Symbolic Computation*, 5, 1988.

[ET75]   S. Even and R. Tarjan. Network flow and testing graph connectivity. *SIAM journal on computing*, 3:507–518, 1975.

[FF62]   L.R. Ford and D.R. Fulkerson. *Flows in Networks*. Princeton Univ. Press, 1962.

[Fud95]   I. Fudos. *Geometric Constraint Solving*. PhD thesis, Purdue University, Dept of Computer Science, 1995.

[Hav91]   T. Havel. Some examples of the use of distances as coordinates for Euclidean geometry. *J. of Symbolic Computation*, 11:579–594, 1991.

[HLS97]   Christoph Hoffmann, Andrew Lomonosov, and Meera Sitharam. Finding solvable subsets of constraint graphs. In *Principles and Practice of Constraint Programming – CP97*, pages 463–477. Springer LNCS 1330, 1997.

[Hsu96]   Ching-Yao Hsu. *Graph-based approach for solving geometric constraint problems*. PhD thesis, University of Utah, Dept. of Comp. Sci., 1996.

[HV94]   Christoph M. Hoffmann and Pamela J. Vermeer. Geometric constraint solving in R^2 and R^3. In D. Z. Du and F. Hwang, editors, *Computing in Euclidean Geometry*. World Scientific Publishing, 1994. second edition.

[HV95]   Christoph M. Hoffmann and Pamela J. Vermeer. A spatial constraint problem. In *Workshop on Computational Kinematics*, France, 1995. INRIA Sophia-Antipolis.

[JC93]   I. Emiris J. Canny. An efficient algorithm for the sparse mixed resultant. In O. Moreno ed.s G. Cohen, T. Mora, editor, *Proc. 10th Intern. Symp. on Applied Algebra, Algebraic Algorithms, an d Error Correcting Codes*, volume 263, pages 89 – 104. Lecture Notes in Computer Science, Springer-Verlag, 1993.

[Kho78]   A.G. Khovanskii. Newton polyhedra and the genus of complete intersections. *Funktsional'nyi Analiz i Ego Prilozheniya*, 12(1):51–61, 1978.

[Law76]   E. Lawler. *Combinatorial optimization, networks and Matroids*. Holt, Rinehart and Winston, 1976.

[Laz81]   D. Lazard. Résolution des systèmes d'équations algébriques. *Theoretical Computer Science*, 15:77–110, 1981.

[Laz91]   D. Lazard. A new method for solving algebraic systems of positive dimension. *Discrete Applied Mathematics*, 33(1):147–160, 1991.

[LM96]   R. Latham and A. Middleditch. Connectivity analysis: a tool for processing geometric constraints. *Computer Aided Design*, 28:917–928, 1996.

[Owe91]   J. Owen. Algebraic solution for geometry from dimensional constraints. In *ACM Symp. Found. of Solid Modeling*, pages 397–407, Austin, Tex, 1991.

[Owe93]   J. Owen. Constraints on simple geometry in two and three dimensions. In *Third SIAM Conference on Geometric Design*. SIAM, November 1993. To appear in Int J of Computational Geometry and Applications.

[Pab93]   J.A. Pabon. Modeling method for sorting dependencies among geometric entities. In *US States Patent 5,251,290*, Oct 1993.

[RAO93]   T.L. Magnanti R.K. Ahuja and J.B. Orlin. *Network Flows*. Prentice-Hall, 1993.

[Ren92]   J. Renegar. On the computational complexity and the first order theory of the reals, part i. *Journal of Symbolic Computation*, 13:255–299, 1992.

[RF96]   O. E. Ruiz and P. M. Ferreira. Algebraic geometry and group theory in geometric constraint satisfaction for computer-aided design and assembly planning. *IIE Transactions on Design and Manufacturing*, 28:281–294, 1996.

[Stu93]   B. Sturmfels. Sparse elimination theory. In *Proc. Computational Algebraic Geometry and Commutative Algebra*, pages 377 –396. Cambridge University Press, 1993.

[Wan93]   D. Wang. An elimination method for polynomial systems. *J. Symbolic Computation*, 16:83–114, 1993.

Desargues: A Constraint-based System for 3D Projective Geometry

Olivier Lhomme[1], Phil Kuzo[1], Pierre Macé[1,2]

We have developed a formalism to simplify the expression of 3D constraints and their solving. Our formalism makes a strong distinction between metric and projective properties. We represent points, lines and planes in projective space by tensors and use Cayley's algebra, with the *join* and *meet* operators, to express projective properties.

We have developed a constraint-based system using this formalism, that both contains knowledge about 3D geometry and algorithms to solve a wide variety of geometric related problems. This formalism is used in cooperation with a known states propagation algorithm.

This work is a part of the *Gina* project ("Géométrie Interactive NAturelle"), aiming at the design of 3D shapes by freehand drawings and oral expressions of properties.

Introduction

Constraint based systems

Constraint based modeling is a growing research domain. However most published works are about 2D or kinematics problems, certainly because of the complexity of constraint solving in 3D.

---

[1]École des Mines de Nantes, La Chantrerie, BP 20722, 44307 Nantes Cedex 3, FRANCE
lhomme@emn.fr, kuzo@emn.fr, mace@emn.fr

[2]Limsi-Cnrs

In fact, in 3D space, the translation of geometrical problems into a set of algebraic equations leads to large systems of equations. So numerical and symbolic methods are not fully satisfactory for solving 3D geometric constraints for cpu time and stability reasons.

For those reasons, the so-called geometric methods are often used: the model is decomposed into classical geometrical problems like curve and surface intersections. All these methods work by propagating symbolic information thanks to either an axiomatization of the problem [Brud93], or to a graph that represents objects and constraints among them [BFH95]. Their limitation is that they solve only simple problems.

The idea for a good modeling system is to couple geometrical and numerical methods in a way to combine the advantages of the two approaches. This idea has been used by a number of some existing systems [BFH95], [BH95] [EHBE96], [Kram92]. Although these systems mix the two main methods - geometrical and numerical - they solve only simple problems and none is fully satisfactory dor 3D constraint solving. Our hypothesis is that this is because of the complexity of constraints in 3D space. Therefore we propose to simplify the expression and solving of these constraints mainly by giving a privileged place to projective geometry.

Gina and Desargues

The aim of the *Gina* project is to build a 3D model of an object knowing a 2D sketch and spatial properties of this object. So the study of the properties preserved by projection (perspective, orthogonal projection...) is of crucial importance. It is the aim of the *projective geometry*.

An analysis of the reasoning of an human expert in projective geometry to design 3D shapes from their 2D sketches let us think that we could derive benefit from the fact that reasoning in projective geometry is often much simpler and intuitive than reasoning in metric geometry.

This led us to design a method that automates the expert reasoning. The method combines a local propagation solver with the inference of some relevant projective properties. It has been implemented in the *Desargues* system. The approach used makes a strong distinction between projective and metric constraints to solve them separately.

The paper focuses on projective constraints only. It is organized as follows:

section 2 recalls some basic concepts about projective geometry. In section 3, we give insight of how the Cayley's algebra can be used to perform computations. Section 4 tracks a typical (human) reasoning in projective geometry, using infinitely distant points. Section 5 presents the *Desargues* system that automates that kind of reasoning.

Projective Geometry

Born with the discovery of perspective and from the work of G. Desargues on the cone, the projective geometry goes over for instance the study of:

- incidence (point on a line, line on a plane...),

- collinearity (points lined up),

- relative positions (point b between a and c...)...

The power of the projective geometry comes from the fact that the infinitely distant points are ordinary points, allowing very intuitive reasonings [X95].

Projective Space

The Euclidean geometry has an algebraic model: the *affine space* in which we represent and compute points according to the Euclid postulate.
To represent points in the projective geometry we use a *projective space*.

Definition

The projective space, *defined on a vector space* V, *is the set* $\mathcal{P}(V)$ *of one dimensional subsets of* V.

If V is a vectorial space of dimension $n + 1$, the dimension of $\mathcal{P}(V)$ is n. The 3D projective space is defined on $V = \mathbb{R}^4$ and each element of $\mathcal{P}(\mathbb{R}^4)$ is called a point and may be represented by one vector of $\mathcal{P}(\mathbb{R}^4)$ or by the product of this vector by any scalar factor. Also it is possible to get this, for instance by homogeneous coordinates.

Homogeneous coordinates

We represent a point of Cartesian coordinates $\begin{pmatrix} X \\ Y \\ Z \end{pmatrix}$ by a vector

$$\begin{pmatrix} x \\ y \\ z \\ w \end{pmatrix} \in \mathbb{R}^4$$

such as (for instance):

$$X = \frac{x}{w}, \; Y = \frac{y}{w}, \; Z = \frac{z}{w}$$

All points with $w = 0$ are infinitely distant points.

Insight of our computation model

The Grassmann-Cayley Algebra gives a formalism for projective space. All *geoms* (points, lines, planes) are represented as tensors. The two operators of this algebra ($\vee$ and $\wedge$) allow the classical operations on geoms (generation, intersection, ...) to be computed. We have developed this in [Macé97].

Cayley Algebra

We call *extensor* an antisymmetric tensor. (The mathematic definitions below are not needed to the understanding of the next sections).

Operator Join
The join *of an m-ordered extensor R and an n-ordered extensor S is the tensor $R \vee S$ of order $m + n$ of which the component with indices I (set of indices) is given [RS76] by:*

$$(R \vee S)_I = \sum_{\substack{J \cup K = I \\ J \cap K = \emptyset}} sign(J, K) \; R_J S_K$$

The sign is given by making the concatenation of the two sets of indices J and K, then counting the permutations of two indices to order these indices: if the number of permutations is even, the sign is $+$ and $-$ in the opposite case.

The *join* is a generalization of the vector product: we combine all components of R and S of which indices may be merged to generate the indices of the component we are computing.

Operator Dual
The dual tensor *of an extensor is the extensor of which the component with indices I is given by*:

$$T_I^* = sign(N - I, I)\, T_{N-I}$$

where N is the set of possible indices and $N - I$ the complement of I in N. We replace a component by the component of which indices are not those of the given component.

Operator Meet
The meet *is the dual operator of the* join.

$$L = P \wedge Q \Longrightarrow L^* = P^* \vee Q^*$$

The *join* is used to generate the sum of two disjoint vector subsets, the *meet* for the intersection and the *dual* to find a supplementary of a vector subset.

This formalism is simple and compact. The rules are the same as applied to lines and planes, to a finitely or an infinitely distant point.

In projective geometry, it is convenient to express all projective constraints in terms of incidence constraints. Incidence constraint systems in a projective space can then be solved in Cayley algebra by the following formulas [Macé97].

Main formulas

- Incidence on a line:

$$a\, \varepsilon\, L \Longleftrightarrow L \vee a = \phi$$

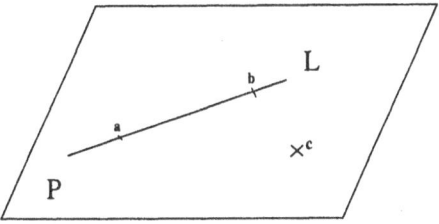

- Incidence on a plane:

$$c \, \varepsilon \, P \Longleftrightarrow P \vee c = 0$$
$$L \, \varepsilon \, P \Longleftrightarrow P \vee L = \phi$$

- Line generation:

$$a \, \varepsilon \, L, \, b \, \varepsilon \, L \Longleftrightarrow L = a \vee b$$

- Plane generation:

$$a \, \varepsilon \, P, \, b \, \varepsilon \, P, \, c \, \varepsilon \, P \Longleftrightarrow P = a \vee b \vee c$$
$$L \, \varepsilon \, P, \, c \, \varepsilon \, P \Longleftrightarrow P = L \vee c$$

- Two planes intersection:

Intersections are generally given by a meet and using duality (but there is a special formula for two intersecting lines [Mour91]). The *alternative laws* [RS76] give us some distributive properties of a great interest to study relative positions (for the canonic base, the brackets are the determinants of vectors components):

$$L = abc \wedge def = [abcd]ef + [abce]fd + [abcf]de$$

- Line/plane intersection:

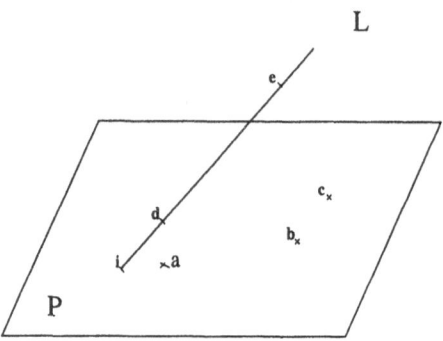

$$
\begin{aligned}
i \;=\;\; & abc \wedge de \;=\; [abce]d - [abcd]e \\
\;=\;\; & [deab]c + [debc]a + [deca]b
\end{aligned}
$$

Before to state the *Desargues* principles, let us track a human expert reasoning for a very simple problem.

Expert Reasoning

We want to reconstruct a 3D shape from a *perspective* 2D sketch. In other words, the 2D sketch is a projection over a projection plane of the 3D shape seen from the eye position.

Reasoning is based - when it is possible - on a conversion of every property in a conjunction of incidences. For instance, let:

- o be the eye,

- Pr the projection plane,

- P_∞ the infinitely distant plane,

the property "to be the projection of" becomes "to be incident to the projection plane and to the line of sight":

$$
a' = proj(a) \iff \begin{cases} \exists \text{ a line L such that } o, a, a'\varepsilon\ L \\ a' \ \varepsilon\ Pr \end{cases}
$$

the parallelism of two lines becomes "to be incident to the same infinitely distant point":

$$L \; // \; M \iff \left\{ \begin{array}{l} i \, \varepsilon \, L \\ i \, \varepsilon \, M \\ i \, \varepsilon \, P_\infty \end{array} \right.$$

Let us show the expert reasoning on the example of a parallelogram (z axis is oriented towards the eye): a 2D sketch $a'b'c'd'$ of a parallelogram $abcbd$ is given. The z-coordinate of point a is also given. The problem is to find the 3D shape of the parallelogram, i.e. the points a, b, c, d.

One can see that the expert applies the two conversions above.

Some steps of the reasoning are skipped by the expert. Remark also that he introduces the plane aij which is the plane containing the parallelogram.

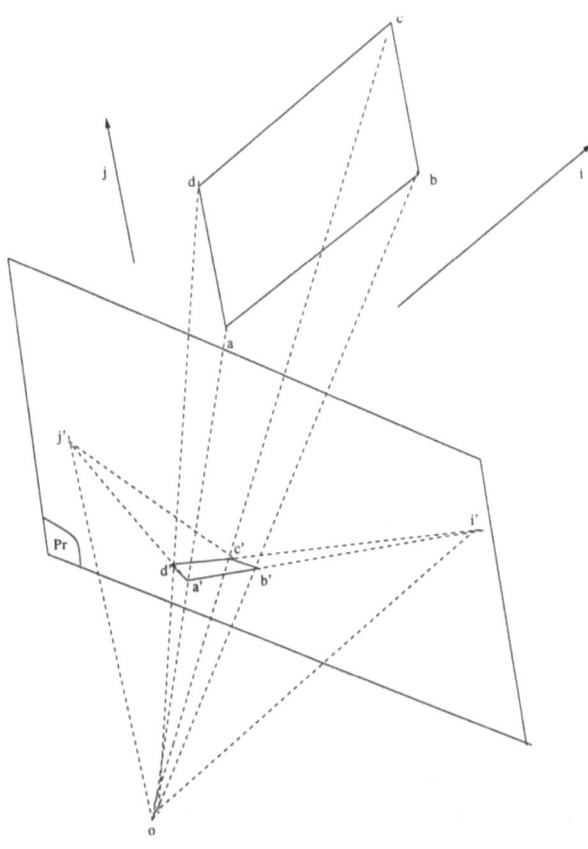

$$\begin{cases} a' = proj(a) \\ b' = proj(b) \\ z(a) = z_0 \end{cases} \quad \begin{cases} ab \ // \ cd \\ bc \ // \ da \end{cases} \quad \begin{cases} i' = a'b' \cap c'd' \\ j' = b'd' \cap d'a' \\ i' = proj(i) \\ j' = proj(j) \end{cases}$$

$$\Downarrow$$

$$\begin{cases} a \ \varepsilon \ oa' \\ b \ \varepsilon \ ob' \\ a \ \varepsilon \ P_{z=z_0} \end{cases} \quad \begin{cases} i \ \varepsilon \ ab \\ i \ \varepsilon \ cd \\ i \ \varepsilon \ P_\infty \end{cases} \quad \begin{cases} j \ \varepsilon \ bc \\ j \ \varepsilon \ da \\ j \ \varepsilon \ P_\infty \end{cases} \quad \{ \ i \ \varepsilon \ oi'$$

$$\Downarrow$$

$$\begin{cases} a = oa' \wedge P_{z=z_0} \\ b \ \varepsilon \ ob' \end{cases} \quad \begin{cases} i = oi' \wedge P_\infty \\ b \ \varepsilon \ aij \end{cases}$$

$$\Downarrow$$

$$b = ob' \wedge aij$$

The *Desargues* system

In this section, we present a method to fully automate the above kind of expert reasoning, which is implemented in the *Desargues* system. The core of the *Desargues* system is a known states propagation algorithm. Local propagation algorithms are incomplete. The inference of relevant projective properties aims at eliminating or at least at reducing this incompleteness.

The *Desargues* system offers a language to help the reconstruction of a 3D shape from a 2D sketch and additional constraints. We focus here only on projective constraints.

Let us outline our approach in very general terms:

1. we introduce a language over *geoms* (points, lines, planes). This language (it will be called the geom-level language) allows 3D geoms to be created and projective constraints between geoms to be stated. The constraint solver is a known states propagation solver (it is not very original and is only briefly described in this paper).

2. we enrich the language with the concept of *bi-geom*: a bi-geom associates a geom with its projection. The bi-geom level of the language allows constraints to be stated over bi-geoms. The bi-geom level of

the language can be implemented on top of the geom-level and boot-strapped, and the presentation given in the paper uses this facility. *Desargues* aims at designing 3D shapes from their 2D sketches, and has specificities from this kind of problem (for example, some geoms always exist in the language: the eye position, the infinite plane and the projection plane). Those specificities of the language allow to infer projective properties to help the known states propagation solver.

3. we show the combination of the known states propagation solver over the geom constraints and the implicit inference mechanism provided by the bi-geom level are sufficient to solve the parallelo-gram problem with a reasoning very close to that of the expert.

Geom-level language

This level makes possible to handle geoms and to express incidence, inter-section, parallelism, collinearity, coplanarity and generation constraints. In fact all those constraints can be defined from a conjunction of incidence constraints (with possibly intermediate geoms).

The *Desargues* system manages three kinds of basic constraints over the geoms:

- type constraints (may be seen as type declarations),

- incidence constraints,

- equality constraints.

A grammar is given below; it will be of use in the next subsection.

```
constraint system :- geom constraint
                      constraint system

geom constraint :- type constraint |
                   incidence constraint |
                   equality constraint

type constraint :- V : T


incidence constraint :- X ε X
```

```
equality constraint :- X = X

X :- V |
     C

V :- geom variable

C :- constant

T :- type constant (line, point, plane)
```

The three basic constraints allow intersection and generation constraints to be easily defined. For example:

- $A \cap B = C$ becomes
  $C \varepsilon A, C \varepsilon B$
  (if the type of the 3 geoms are known)

- *P generated by* A, B, C becomes
  $A \varepsilon P, B \varepsilon P, C \varepsilon P$
  (if A, B, C are points and P is a plane).

A classical known states propagation algorithm tries to solve the constraint system: whenever the degree of freedom of a geom is 0, *Desargues* tries to combine the incidence constraints linking that geom to determine its value.

Bi-geom level language

The key idea is rooted in the definition of a projective constraint itself: a projective constraint is preserved by projection. Then every projective constraint between some geoms of the 3D model induces a constraint between the projections of the geoms. The method will consist in always inferring those induced constraints. The system creates and handles for each 3D item (geom or projective constraint) the projected item that corresponds to it.

To explain the method more precisely, we need three geoms that always exist: the eye position (E_{ye}), the infinite plane (P_∞)and the projection plane (P_p). They are geoms like others: they may be unknown and constrained, but their names are given in the language.

Notation: if X is a bi-geom, X_{3D} and X_p respectively denote the geom in the 3D model and its projection.

The *Desargues* language is then extended to handle bi-geom. We will not give a grammar of the extended language. We will just use the following convention: we differentiate a constraint symbol working on bi-geoms from the constraint symbol working on geoms by simply doubling the latter (e.g., $\varepsilon\varepsilon$ denotes the incidence constraint over bi-geoms).

The next subsections give examples of how primitives over the bi-geoms can be defined. They are defined in the *Desargues* language.

Creation of a bi-geom X of type T

The creation of a bi-geom X of type T, denoted

$$X :: T,$$

consists in creating two geoms and stating the constraint that one is the projection of the other. This can be done by the following *Desargues* code:

$X_{3D} : T$
$X_p : T$
$Ray : line$
$X_{3D} \ \varepsilon \ Ray$
$X_p \ \varepsilon \ Ray$
$E_{ye} \ \varepsilon \ Ray$
$X_p \ \varepsilon \ P_p$

X_p could be of a dimension smaller than X_{3D}, e.g. in the case of a line parallel to the direction of the projection, but for simplicity we do not discuss this point.

Ray denotes the line of sight: X_{3D}, X_p and E_{ye} are all on that line.

Incidence constraint between two bi-geoms X and Y

The corresponding *Desargues* code for $X \ \varepsilon\varepsilon \ Y$ is given below: it consists in stating two incidence constraints, one over the 3D geoms, one over the projections of the 3D geoms.

$X_{3D} \ \varepsilon \ Y_{3D}$
$X_p \ \varepsilon \ Y_p$

Parallelism constraint between two bi-geoms X and Y

The parallelism constraint can also be defined from geoms and geoms constraints, but the definition is more concise with the use of an intermediate bi-geom. PP_∞ denotes here the bi-geom corresponding to the infinite plane P_∞. If two geoms are parallel, their bi-geoms X and Y verify:

I :: point
I εε PP_∞
I εε X
I εε Y

Desargues reasoning

Let the eye position and the infinitely distant plane be known, let us give the following program to *Desargues*:

```
% definition of a parallelogram
(1)  A,B,C,D::point
(2)  AB::line(A,B)
(3)  BC::line(B,C)
(4)  CD::line(C,D)
(5)  DA::line(D,A)
(6)  AB // CD
(7)  BC // DA
% data from the 2D sketch
(8)  Ap = value-of-projection-of-A
(9)  Bp = value-of-projection-of-B
(10) Cp = value-of-projection-of-C
(11) Dp = value-of-projection-of-D
% additionnal data giving a z-coordinate to A
% (otherwise the problem is underconstrained)
(12) PlaneZA ::  plane
(13) PlaneZA == value-of-a-z-plane
(14) A εε PlaneZA
```

The "reasoning" followed by *Desargues* is simply a known states propagation, but combined with the induced constraints. It is very close of the one followed by the expert:

- A is determined by its intersection with PlaneZA and the line of sight joining the eye, A_p and A (constraints 8 and 14).

- Since AB and CD are parallel, the intersection I of AB and CD is in the infinitely distant plane (constraint 6 defines such a bi-geom I, see 5.2.3). Its projection I_p is the intersection of $A_p B_p$ and $C_p D_p$ and can be computed.

- From I_p *Desargues* get I by the intersection of the line of sight eye/I_p and the infinitely distant plane.

- B is at the intersection of the line AI and the line of sight eye/B_p

- and so on for C and D

In fact that reasoning is more economic that the one followed by the expert: *Desargues* does not introduce the plane of the parallelogram.

We have successfully used *Desargues* to reconstruct various polyhedrons from their 2D sketches (hexaedron and parallelepiped).

Future work

Although the method is very simple, the *Desargues* system is a powerful tool for designing 3D shapes from their 2D sketches, in particular in the domain of architecture where projective constraints are the more frequent ones.
Desargues is being extended to handle non-projective constraints by incorporating geometric methods like the ones used in [BFH95, BH95, Kram92] as well as numerical methods.

Acknowledgments

Thanks to Jean-Daniel Fekete for his helpful comments on the paper.

References

[BFH95]   W. Bouma, I. Fudos, C. Hoffmann, J. Cai, R. Paige. A ge-
          ometric Constraint Solver. Computer Aided Design, 27(6), pp
          487-501, juin 1995.

[BH95]   B. Brüderlin, C. Hsu. A degree of freedom graph approach
          to solving geometric constraint problems. Theory and prac-
          tice of geometric modelling (Blaubeuren II). To be published
          by Springer Verlag 1997.

[Brud93] B. Brüderlin. Using Geometric Rewrite Rules for Solving Geo-
          metric Problems Symbolically. Theoretical Computer Science,
          116(1993), pp. 291-303, Elsevier.

[EHBE96] L. Eggli, C-Y. Hsu, B. D. Brüderlin and G. Elbert. In-
          ferring 3D Models From Freehand Skethes And Constraints,
          Computer-aided Design, Vol. 29, pp. 101-112. Copyright 1996
          Elsevier Science Ltd.

[Kram92] G. Kramer. A Geometric constraint engine Artificial Intelli-
          gence, 58 (1992), pp 327-360 Elsevier.

[Macé97] P. Macé. Tensorial Calculus of Line and Planes in Homoge-
          neous Coordinates, Computer Network and ISDN Systems 29
          (1997) 1695-1704, Elsevier Science B.V.
          http : //www.emn.fr/dept_info/perso/mace

[Mour91] B. Mourrain. Approche effective de la théorie des invariants
          des groupes classiques. Thèse Ecole Polytechnique, septembre
          91.

[RS76]   G.-C. Rota et J. Stein. Application of Cayley Algebras Collo-
          quio Internationale, Roma 1973,Tome II,pp 71-97,Attidei Con-
          veigni Lincei *n°* 17,pub. Accad. Naz. Lincei, Roma 1976.

[X95]   "Rupture Culturelle" en Simulation Informatique. Technolo-
          gies 'France', juin 1995.

YAMS: A Multi-Agent System for 2D Constraint Solving

P. Mathis, P. Schreck, J.-F. Dufourd[1]

In spite of recent trends, geometric constraint solving remains an important topic in CAD. The first methods were mainly based on algebraic or geometric approaches. For several years, current methods, proceed by decomposition of the problem into sub-problems easier to solve. Recently, we have exposed a general formalization of the principles of these approaches and showed that they are all based on the invariance under displacements of the CAD constraint systems.

Concretely, we implemented these ideas and made a solver which is able to integrate different and independent solving methods. This solver works in two phases: first, a formal one which produces a plan of construction, and then, an interpretative one which produces numeric and graphical solutions.

This paper describes our solver as a multi-agent system where each agent implements a local solving method. The first two agents are geometric knowledge-based systems and the third is an algebraic one. The agents produce some parts of the construction plan which are next assembled by a kind of super-agent, so-called supervisor. Local solvers are activated according to their priority and their ability to solve a part of the problem.

Introduction

In Computer-Aided Design (CAD), the automatic construction of an object defined by a *system of geometric constraints* remains a topical sub-

---

[1]LSIIT, Université Louis Pasteur,
7, rue René Descartes, 67084 Strasbourg, France

ject even in 2D. Such an object is usually defined through a dimensioned sketch specifying the metric relationships and the topological constraints like incidence and adjacency relationships between its components. The sketch is completed with dimensional constraints interactively entered by the designer. The problem is then to build an exact figure satisfying all these constraints.

In this kind of problem, the metric relationships are often numerically given and a numerical solving is enough to satisfy the designer. But sometimes a formal —or symbolic— solving is needed, for instance in Computer-Aided Instruction (CAI) or in CAD when some dimensions are symbolically given. We think that there are significant advantages to considering a formal solving. For instance, this way keeps the possibility of producing several numerical solutions, it eliminates convergence problems and favors the diagnosis of failures. All these possibilities come from the fact that a formal solution is an exact solution. Another great advantage lies in the generic character of the solution. Indeed, a formal solving consists in two phases : in the first phase a construction plan is given (if necessary, the numeric values of the given dimensions are temporarily abstracted) and in the second phase, this plan is interpreted with the numerical values of the parameters to yield numerical and graphical solutions. This way, when the numerical values of some abstracted dimensions are changed, the system must only run the second phase which is the lighter one.

Recent works [Owe91, BFH+95, LM95] use the notion of decomposition of a geometric construction problem mainly to extend the solving power of the construction process. As said in [DMS97], the nature of these decomposition algorithms is geometric and comes from the *invariance under displacements* of the CAD construction problems. Showing clearly this phenomenon, we have presented in [DMS98] an approach which make compatible decomposition and formal solving. Moreover, our approach makes it possible to use different local solving methods and so to combine their solving power. These ideas have been implemented as a *multi-agent system with blackboard*. In this architecture, each agent implements a local solving method which is able to solve a part of the located constraint system. These agents are launched and controlled by a super-agent called the *supervisor*. The supervisor makes the construction progress deciding to *assemble* the solved subsystem which have enough common information. The agents work with a common database, usually called a blackboard, which contains the constraints and the solved subsystems.

The paper is structured as follows. First, we expose the theoretical foundation of our approach. Then, we present our multi-agent system whose agents are detailed in the next section. Before ending we show some significant examples and we give some conclusions.

Formal solving using decomposition modulo displacements

We use the usual terminology of the equation systems. But we add some logical considerations to handle our formal geometric framework.

Geometric universe

The terms describing the constraints come from a geometric universe whose preferential interpretation is the Euclidean plane with a coordinate system $(O, \vec{\imath}, \vec{\jmath})$ fixed once and for all.

The syntactical description of this universe is given by a *heterogeneous signature* Σ containing a set Θ of *type* symbols. In Θ, *Point, Line, Circle, Length, Angle, Conic, Displacement* and *Reference* are geometric types and *Real* is a numerical type. Each element of type α of Θ is interpreted as a set E_α of objects with type α. We assume that a *degree of freedom*, linked to the notion of coordinates, is associated with each type α. Signature Σ contains other symbols, namely *functional* and *predicative* symbols. These symbols are characterized by a profile indicating the type of their arguments and result. They are interpreted by functions and predicates on sets E_α. So, the geometric universe can be seen as a universe of terms in which predicative terms express constraints and functional terms express constructions.

Figure, constraint system

We define a *figure* as a n-tuple —we say a *vector*— $f = (o_1, \ldots, o_n)$ of geometric objects of respective types $\alpha_1, \ldots, \alpha_n$. We say that $f' = (o_{i_1}, \ldots, o_{i_k})$ is a *subfigure* of f if it is one of its subvectors. A formal or *parametric* figure is a n-tuple of functional terms or constants of the geometrical universe.

The geometric figures can be described declaratively by a *system of geometric constraints, constraint system* in short. More precisely, a constraint system S is a triplet (X, A, C) where X is a set of *unknowns*, de-

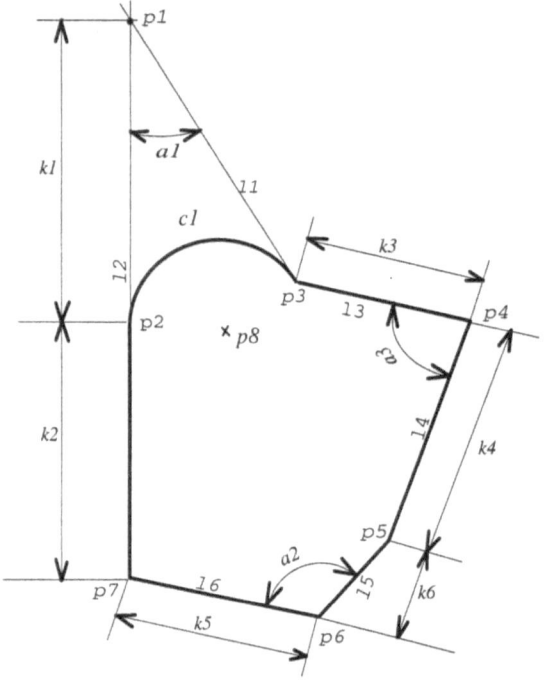

Figure 1: A constraint system

noted by $\mathcal{I}(S)$, A a set of *parameters*, denoted by $\mathcal{A}(S)$, and C a set of *constraints*, denoted by $\mathcal{C}(S)$, of the form

$$C = \{p_1[X, A] \ldots, p_m[X, A]\}$$

where each $p_i[X, A]$ is a predicative term, namely a *constraint* whose variables are in X or in A. Unknowns from X and parameters from A are considered with their types. For instance, the dimensioned sketch in Figure 1 can be textually described by the system whose constraints are given on Table 1. In this constraint system, parameters are the symbols $\{k1, \ldots, k6, a1, \ldots, a3\}$, where ki are lengths and ai are angles, and the unknowns are the symbols $\{p1, \ldots, p8, l1, \ldots, l6, c1\}$, where pi are points, li are lines, and $c1$ is a circle. Note that some constraints are not represented on the sketch such as tangency and the explicitly given constraint $fixplv(p7, l2)$ meaning that the point $p7$ is fixed at the origin O of the coordinate system and the line $l2$ has a vertical direction.

Table 1: Constraint system of the Figure 1

onl(p1, l1)	% point p1 is on line l1	fixplv(p7, l2)	% point p7 is fixed and
onl(p1, l2)			line l2 has a vertical direction
onl(p2, l2)		distpp(p1, p2, k1)	% the distance from p1 to p2 is k1
onl(p3, l1)		distpp(p7, p2, k2)	
onl(p3, l3)		distpp(p4, p3, k3)	
onl(p4, l3)		distpp(p5, p4, k4)	
onl(p4, l4)		distpp(p7, p6, k5)	
onl(p5, l4)		distpl(p5, l6, k6)	
onl(p5, l5)		angle(p1, p2, p1, p3, a1)	
onl(p6, l5)			% the angle of lines p1p2, p1p3 is a1
onl(p6, l6)		angle(p6, p5, p6, p7, a2)	
onl(p7, l2)		angle(p4, p3, p4, p5, a3)	
onl(p7, l6)		tgclp(c1, l1, p3)	% circle c1 is tangent to line l1 at p3
onc(p2, c1)	% point p2 is on circle c1	tgclp(c1, l2, p2)	
onc(p3, c1)		centre(c1, p8)	% p8 is the center of circle c1

Solution, triangular form

A solution is a valuation from the set of the unknowns to a model of the geometric universe. While a numerical solution, if it exists, can be seen as a figure, a formal solution can be seen as a parametric figure. When the system S contains no parameter, we note $\mathcal{F}(S)$ the set of figures whose elements are solutions for S. If $\mathcal{F}(S)$ is finite and nonempty, we say that S is *well-constrained*. If it is empty, we say that S is *over-constrained*. If it is infinite, we say that S is *under-constrained*. When S contains parameters, we have two sorts of solutions : formal solutions and their interpretations. We consider, here, that there is only one solution which is the function computing for each element a of the parameter space, the set $\mathcal{F}(S_a)$ of the solutions for the corresponding system S_a where parameters are instantiated by values of a.

Thus, formal solving consists in producing a parametric figure interpreted as the function $a \mapsto \mathcal{F}(S_a)$. Following the equation system terminology, this solving performs the triangulation of the constraint system to reduce it into *triangular solved forms*. Such a form is a particular kind of constraint system where all the constraints $p_i(X, A)$ are under the form

$$x_i = f_i[x_1, \ldots, x_{i-1}, A]$$

where $f_i[x_1, \ldots, x_{i-1}, A]$ denotes a functional term containing only the unknowns $x_1, \ldots, x_{i-1}$ and parameters of A. So, we say that a parametric system S is *solvable* if there is a set T of $m \geq 1$ solved triangular systems $T_1, \ldots, T_m$ with $\mathcal{A}(T_i) = \mathcal{A}(S)$, $\mathcal{I}(T_i) \supseteq \mathcal{I}(S)$ and T is equivalent to S. $\mathcal{I}(T_i)$ may be bigger than $\mathcal{I}(S)$ because of the use of auxiliary elements like lines, circles which are not in the sketch. If $\mathcal{F}(T_a) \subset \mathcal{F}(S_a)$ for each value a of the parameter space, then the resolution is *incomplete*. If

$\mathcal{F}(S_a) \subset \mathcal{F}(T_a)$, then the resolution is *incorrect*.

For instance, the constraint system of the example on Table 1 is solvable in our geometric universe. Our prototype *YAMS* gives the formal solution of the Table 2. As explained in [DMS97], this single solved system represents in fact all the triangular systems $T_1, \ldots, T_m$ interpreting the functional symbols as *multifunctions*. This is just an illustrative example: some notions used will be explained further, and the complete construction done in greater details in section dealing with examples.

Table 2: Construction for example on Table 1

```
% fixing of the metric parameters
    k5 = initl(250.000)
    k4 = initl(158.000)
    a3 = inita(1.413717)
    k3 = initl(304.000)
    a2 = inita(2.809980)
    k2 = initl(200.000)
    k1 = initl(300.000)
    a1 = inita(0.558505)
    k6 = initl(70.000)
% subplan suffixed by .1
    p7.1 = initp(-296.958, -248.927)
    l2.1 = initd(p7.1, 1.561722)
    a12 = inv_angle(supp_angle(a1))
    c3.1 = mkcir(p7.1, k2)
    p2.1 = interlc(l2.1, c3.1)
    c4.1 = mkcir(p2.1, k1)
    p1.1 = interlc(l2.1, c4.1)
    a7.1 = inita(1.570796)
    l7.1 = lpla(p2.1, l2.1, a7.1) 1
    a9.1 = bissect(a1)
    l8.1 = lpla(p1.1, l2.1, a9.1)
    p8.1 = interll(l8.1, l7.1)
    l1.1 = lppa2(p1.1, p2.1, p1.1, a12)
    c1.1 = mkcir2(p8.1, p2.1)
    p3.1 = interlc(l1.1, c1.1)
    a55 = inv_angle(a3)
% subplan suffixed by .2
    p5.2 = initp(0.000, 0.000)
    l4.2 = initd(p5.2, 0.000)
    c6.2 = mkcir(p5.2, k4)

    p4.2 = interlc(l4.2, c6.2)
    c7.2 = mkcir(p4.2, k3)
    l3.2 = lpla(p4.2, l4.2, a55)
    p3.2 = interlc(l3.2, c7.2)
    a105 = inv_angle(a2)
% subplan suffixed by .3
    p7.3 = initp(0.000, 0.000)
    l6.3 = initd(p7.3, 0.000)
    l9.3 = ldl(l6.3, k6)
    c11.3 = mkcir(p7.3, k5)
    p6.3 = interlc(l6.3, c11.3)
    l5.3 = lpla(p6.3, l6.3, a105)
    p5.3 = interll(l5.3, l9.3)
    k9 = fdist(p7.1, p3.1)
    k10 = fdist(p5.2, p3.2)
    c15.3 = mkcir(p7.3, k9)
    c16.3 = mkcir(p5.3, k10)
    p3.3 = intercc(c16.3, c15.3)
% definition of the displacement dep1.3 from 2 to 3
    dep1.3 = make_dep_pp(p3.3, p5.3, p3.2, p5.2)
% use of displacement dep1.3
    p4.3 = transfp(p4.2, dep1.3)
% now the subplan .1 is completed
% definition of the displacement dep2.1 from 3 to 1
    dep2.1 = make_dep_pp(p3.1, p7.1, p3.3, p7.3)
    p6.1 = transfp(p6.3, dep2.1)
    p5.1 = transfp(p5.3, dep2.1)
    p4.1 = transfp(p4.3, dep2.1)
```

Invariance under displacements

Recall that the displacement group is the group of punctual affine transformations preserving distances and orientations. This is the reason why the displacements are also named direct isometries or rigid body motions. In the plane, the displacement group only contains translations and rotations. One of the most significant characteristics of the figures specified by a system of measurements in CAD is their independence with respect to any coordinate system. In other words, the constraint systems considered in CAD are invariant under the group of displacements.

The property of invariance can be explained more deeply by the action of the displacement group over the geometric universe. So, the displacement group acts over the types, the functions and the predicates. The action of displacements can be extended classically to the other geometric types. Some types, for instance *Length* or *Angle*, are peculiar: the displacements have no effect on them. We called them *metric types*. There are other interesting kinds of geometric objects: those for which, given two objects o and o' of the same type, there is one and only one displacement φ such that $o = \varphi(o')$. We say that these types define *references*: a classical example is the product type *Point* × *Direction* which is so typical that we call it the *Reference* type.

Displacements have also some effects on the functions. In particular, a function g such that, for each displacement φ, holds the equality $g(x_1, ..., x_n) = g(\varphi(x_1), ..., \varphi(x_n))$ is so called *stable under displacements*. Likewise, a constraint of the form $p[X, A]$ is *invariant under displacements* if, for each displacement φ, $p[\varphi(X), A]$ is equivalent to $p[X, A]$ and a constraint system S is *invariant under displacements* if each constraint of $\mathcal{C}(S)$ is invariant under displacements.

The invariance under displacements of a constraint system S gives a particular form to the solutions set : it is invariant under displacements, more precisely, it is a union of orbits of the displacement actions. If the number of such orbits is finite non null, we say that S is well-constraint *modisp*. Each orbit solution is called a figure *modisp* solution of S.

Since there is an infinity of solutions for an invariant system under displacements, it is indispensable to add to it some constraints to particularize one figure in an orbit. A constraint fixing a reference in S is called in one blow a *reference constraint*. Then, if S is well-constraint *modisp*, the system $S + r$ obtained by addition of the reference constraint r in S, is well-constrained and can be classically solved. The system $S + r$ is said to be located. With the example described by Figure 1 and Table 1, one can pass from a located constraint system to an invariant system forgetting the constraint *fixplv(p7, l2)*. Then, the solutions can be placed anywhere in the plane: the set of the figures obtained from one particular solution by all the displacements is a solution *modisp* of S. Details of these theoretical aspects are in [DMS97, DMS98].

Partial solving and assembling

Unfortunately, a bad choice for a reference can lead to a difficult construction problem while another choice would produce a simple problem. In fact, the situation is worse: it may happen that, even if the system is well-constrained *modisp*, the resulting construction problem is only partially solved whatever the chosen reference. In our example, fixing points $p7$ and $p2$ on line $l2$ enables to obtain a solution *modisp* only for the unknowns $p1, p2, p3, p7, l1, l2$ and $c1$ (see Figure 5).

Some properties of the constraint systems invariant under displacements can be used to overcome this problem. First, a subsystem of an invariant system is itself invariant under displacements. Second, a reference constraint can be relaxed using a displacement. Precisely, for two distinct reference constraints r and r', fixing different unknowns, there exists, for any solution of $S + r$, a displacement φ moving it into a solution of $S + r'$. So, when a constraint system is partially solved using a reference constraint, we can attempt to solve the remaining subsystem using another reference constraint and then, to *assemble* the two partial solutions with the appropriate displacement. In our example, the entire solution requires three local solvings characterized in Table 2 by the suffixes 1, 2 and 3. At the end, displacements are defined and used to translate and to assemble subsystems 2, 3 and 1 in the first reference.

Since the already solved part gives only particular solutions linked to the choice of the first reference, we cannot use the values of the solved unknowns to continue the solving with *another* reference constraint. The unknowns of the remaining part must be considered as unknowns even if they are solved elsewhere. So, the remaining part is often underconstrained. Once again, the invariance under displacements allows to build new invariant constraints increasing the residual subsystem to make it well-constrained. These constraints express the metric relationship of any *subfigure* solution of the solved subsystem and they make up an invariant constraint system that we call the *border* of the solved subsystem. For instance, in our example (see Figure 5), the distance named $k9 = fdist(p3.1, p7.1)$ is the same whatever a reference is chosen. So the term $distpp(p3.3, p7.3, k9)$ is a constraint of the border of subsystem 3.

When the common unknowns of two subsystems, solved each of both with a particular reference constraint, define at least a reference — for instance one point and one direction, or two points — we say that the two subsystems have a *common reference*. If so, it is possible to merge the two subsystems to produce the one of the global system. This operation is the assembling process, in short the *assembling*. It uses displacements to "glue together" two partial solutions. We have shown that this operation is correct and complete [DMS98].

A multi-agent system

General presentation

The topological modeling tool named *Topofil* [BDFL93, BD94] has been extended to include the ability to solve constraint systems. This new prototype is called $YAMS^2$. The interface of the modeling tool, used to draw sketches, has been augmented in order to enable the user to enter metric constraints. Topological constraints, i.e. incidence and adjacency constraints, are deduced from the sketches.

$YAMS$ formally solves the geometric constraint systems by assembling subfigures obtained by application of different local solving methods from a fixed initial local reference. The formal solver has the shape of a *multi-agent system with a blackboard* [WM95]. The architecture of this kind of system has three components (see Figure 2). The blackboard is a shared memory which, in our case, contains the constraint system and partial solved subsystems. The agents deal with data blackboard. In our framework, they implement the solving methods. Presently, $YAMS$ owns two kinds of agents: formal knowledge-based systems directly working on the geometric constraint system, and an algebraic/numerical agent also working with a translation of this system into a polynomial equation system. The supervisor is a peculiar agent which triggers the other agents and updates the blackboard. The result of the multi-agent system is a formal construction plan in the sense of the previous section.

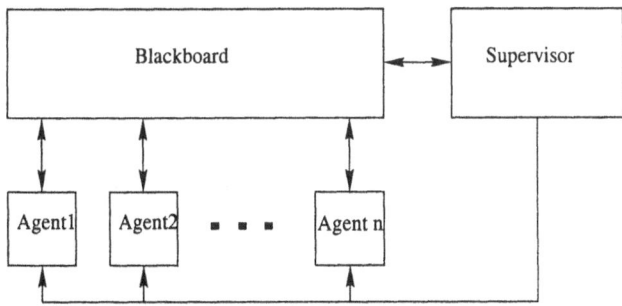

Figure 2: Multi-agent architecture

---

[2]Yet Another Meta Solver

The construction plan is numerically evaluated in an interpretative phase. The result is returned to the modeling tool. One can note here that the set of numerical solutions is often bigger that expected. On the one hand, wrong solutions can appear because of the use of strict implications by some local solving methods such as our rule based systems containing rules weaking the knowledge. A verification of constraint satisfaction is needed to filter the results. On the other hand, solutions whose shape is far from the sketch are yielded by the solver and not wanted by the user. Also, heuristics are used in order to narrow down the set of solutions. Some studies are in progress in our team to efficiently explore the space of solutions.

Like the modeling tool *Topofil*, the prototype was developed in C language for Silicon Graphics workstations using the GL graphic library. The next section presents the architecture of *YAMS*. More technical details are given in [Mat97].

Blackboard architecture

From a theoretical point of view, the blackboard implements a first order logical theory with equality. From a practical point of view, it manages a set of terms corresponding to the problem statement, the intermediate facts and the solved subsystems. Initially, the blackboard contains only the constraint system corresponding to the statement. Finally, if the resolution succeeds, it contains the plan of construction as well. Each local solving method accesses and updates the blackboard. The supervisor — which controls the blackboard— examines it at each step to detect the subsystems having a common reference and to assemble them.

Geometrical universe

As indicated above, our geometric universe is described by a signature and some axioms. The signature contains the types *Point, Line, Circle, Angle, Length, Displacement, Reference* and *Conic*. Note that the last three types, are used only for auxiliary objects. For instance, *Conics* are used to solve tangency constraints. The signature also contains functional symbols, like *mid* (midpoint), *line, center*, to express the construction plans, and predicative symbols, like *distpp* (distance between two points), *angle, distpl* (distance between a point and a right line) to express the constraints used by the inference mechanism.

In this framework, two types of axioms are considered: the facts, which are closed predicative terms corresponding to the initial constraints —we say that they particularize the geometric universe—, and geometric rules, which express some general geometric properties not directly linked to the construction process. Theorems are derived from these axioms with the help of two classic inference rules, namely the *paramodulation* rule, which takes in account the equalities, and the *modus ponens*.

The set of facts implements the current constraint system. There are two possible forms for these facts: plain constraints $P(e_1, ..., e_n)$ and definitions $e = f(e_1, ..., e_n)$ expressing one explicit construction step, where P is a predicative symbol distinct of the equality symbol =, f is a functional symbol, and arguments e_i are logical constants.

The general geometric rules formalize some properties involving the equality, such as the permutability of the arguments of some constraints, or the Chasles relation, or other geometric properties expressible by an implicative form $p_1 \land ... \land p_n \supset q$. These rules define the general framework of our geometric domain: this knowledge is destined to be completed by the specific construction rules used by additional system expert agents.

Since we single out the Euclidean space $I\!R^2$ for model, we add some semantic notions to the syntactic universe. Some are linked to each geometric type, like degree of freedom and degree *modisp*, others are linked to each functional symbol like interpretation of the multifunctions or, linked to each predicative symbol like restriction degree, polynomial equation. For instance, the circle has 3 degrees of freedom, 2 degrees modisp — one degree is eliminated intuitively because the radius is invariant under displacement —, the *intercc* function symbol is linked to a numerical function computing the intersection between two circles, the *distpp* predicative symbol is linked to the classical implicit equation of the distance between two points.

We detail now our implementation of the geometric universe into the blackboard.

Terms management

General treatment   From the external point of view, the blackboard is a set of subsystems and a set of terms, i.e. facts and definitions, from which the agents and the supervisor extract useful information. Agents and the supervisor can, in turn, add new facts to this database. But, internally, facts are not managed in the same way. Most of the facts are explicitly represented, however, we use special data structures for facts having peculiar properties like permutation in distance constraints, Chasles relation for angle constraints and so on.

Furthermore, the number of constraints of a border has an exponential growth according to the number of border unknowns. While in practice there are few unknowns in the border of a subsystem, we avoid the pre-computation of the border constraints. To satisfy agents requirements, specific procedures extract the constraints of the border without storing them in the blackboard. Thus, in our prototype, the border is only implicitly represented. We now explain some particularities of our terms management.

Example of two peculiar cases: equality and Chasles relation Equality and angle constraints generate two equivalence relations. We use special arborescent data structures to manage the resulting equivalence classes [Mat97]. In the case of the angle constraints, we have done some developments to keep the angle values in the classes. In the case of the equality, a particular constant is distinguished and represents all the other constants of the class: this representative constant is used by the unification mechanism. The detection of a new equality or a new angle constraint by the solving agents causes the updating of the corresponding equivalence classes i.e. the fusion of the corresponding trees and the computation of the new labels.

Located solved subsystem During the building of the construction plan, several subsystems are considered, each located with its own reference constraint. These solved located subsystems are exploited in three different ways :

- first, an agent can complete them by solving new constraints;
- second, the border informations are used to solve other subsystems;
- third, common references between two subsystems are searched for the assembling process.

To make the data access easier, we have made concrete the notion of solved subsystem in our blackboard architecture. Thus, the blackboard procedures maintain a graph structure whose nodes are the already located solved subsystems and edges represent common unknowns between two subsystems. This way, the supervisor can quickly compute the common references and assemble the subsystems as soon as possible.

Border The border is a particular constraint system computed from a solved subsystem. A solved subsystem contains the definitions of elements such as points, lines, circles and so on. Elements of a subsystem S are unknowns of its border of S if they are constrained with elements which are not in S. The constraints of the border are formed from the metric relations (distance, angle, etc.) between the border elements [DMS98].

In the border, only the elements explicitly present in the sketch are relevant. So, the unknowns of the border are points, lines and circles. In our geometrical universe, the metric relations are distances between two points, distances between a point and a line, and angles between two lines. The invariant part of an object under the displacements is not regarded, so, in the case of circles on the border, only their center is considered.

Supervisor

The supervisor has many functions. It starts up agents, assembles the subsystems, orders the borders unknowns evaluation, filters the data between agents and blackboard and, finally, stops the formal solving process.

The agents produce subsystems which contain at least one reference, for instance a point and an incident line but never a single point or a single line. There is a priority for each of our three agents. First of all, the supervisor triggers the agent with the highest priority. This agent begins to generate definitions corresponding to a reference. For instance, if point p_1 and line l_1 are incident, they can be chosen and the agent adds to the blackboard the definitions $p_1 = initp(0.0, 0.0)$ and $l_1 = initl(p_1, 0.0)$. Next, the agent solves other constraints as long as it can. All the definitions produced constitute a new subsystem. The supervisor starts the same agent again as long as it yields new solved subsystems. Then, the supervisor triggers the next agent according to the priorities. If this second agent produces definitions, the supervisor calls anew the highest priority agent, else it calls the next agent. Each time an agent begins a new solving, it chooses a never used reference or tries to continue the construction of a prior subsystem which can be increased due to the border information of others prior subsystems.

Before triggering an agent, the supervisor makes all possible assemblings. The supervisor consults the adjacency graph of solved subsystems to assemble them. Each time two subsystems have at least one common reference, they are assembled. We have proved that when several subsystems can be simultaneously assembled the order to assemble them does not matter [DMS98].

Agents

Geometrical agents

Some local solving methods are formal geometric methods in the way of Aldefeld or Brüderlin [Ald88, Bru93]. Actually *YAMS* works with two such agents. They are based on the same logical theory —which is an extension of the blackboard theory— and differ from each other by their solving strategy. Both use the *loci method* [Pet90, Leb50, But79, Sch94a, Sch94b, DS94] and infer new relations from initial constraints. The deduced relations are either *defining* incidence relations giving geometric loci yielding to the construction of geometric objects or new constraints.

As mentioned above, the axioms of our logic theory have the form either of closed atoms or of implications with variables. Variables involved in the premises are always universally quantified. The others, only present in the conclusions of implications, are existentially quantified. There is a single inference rule: the *modus ponens*. Both of the knowledge-based systems contain *production rules* corresponding to the implications of the logical theory. Each production rule has the form

$$\text{if } P_1, \ldots, P_n \text{ then } C_1, \ldots, C_m \text{ conditions } D_1, \ldots, D_p$$

where P_i are premises, C_i are conclusions, and D_i are extra-logic conditions. The conclusions are new facts or definitions like

$$x = f(x_1, \ldots, x_k)$$

where f is a function symbol, the x_i and x are variables. The construction plan increases in size each time such a definition is produced. If x is existentially quantified and replaced by an unknown of the unsolved constraint system, then the functional term defines the unknown.

Conditions in the production rules do not modify the meaning of the geometric construction. They avoid a proliferation of irrelevant relations. Indeed, the verification of conditions helps to test whether objects are already defined or not in the local reference. Applying a rule yields new relations only if the constants substituted for the variables validate the conditions.

For instance, the following rule extracted from our knowledge base expresses that, if the distance L between two points X and Y is known and Y is not defined yet and X is defined — implicit condition — then Y is on a circle C defined by $C = mkcir(X, L)$ and the center of C is X.

```
if distpp(X, Y, L)
```

```
then onc(Y, C), C = mkcir(X, L), centre(C, X)
conditions undef(Y)
```

The first knowledge based-system, which has the highest priority, is based on a strategy usable when there is no loop risk. Thus, the first system only contains production rules whose conclusions never match the premises, and uses a *depth first search*, which is the most efficient strategy for this geometric framework. Such production rules are convenient in most of the CAD problems. For instance, if the rule above is triggered, this strategy tries first to apply rules with the new facts i.e. incidence on a circle and the center.

However, the need for a more complete module of geometric construction prompted us to consider in our second knowledge-based system more complex rules which often produce loops. A *breadth first search* with a limitation on the depth of the search tree is then used. It is not complete, but succeeds in most cases in achieving sophisticated constructions.

Algebraic agent

The algebraic agent makes a polynomial system to be solved by the Newton-Raphson method [Mor83, PTWF92]. This numerical method is very useful in CAD softwares because it can solve most of the constraints. However, because of its drawbacks, we prefer other methods when it is possible. So, the algebraic agent is triggered to continue a solving when the two knowledge-based systems fail.

In the first formal phase, the algebraic agent prepares a minimal geometric constraint system which has to be solved. A hidden corresponding numerical equation system is then generated and numerically solved by the Newton-Raphson method in the second interpretative phase. A minimal and geometric constraint system is written.
$$\{p_1[X, A], \ldots, p_k[X, A]\}$$
where $X = \{x_1, \ldots, x_n\}$ is the set of geometric unknowns and the $p_i[X, A]$ are the constraints at work. This system is used to produce the global definition
$$x_1, \ldots, x_n = NR(p_1, \ldots, p_k, x0_1, \ldots, x0_n)$$
where $x0_1, \ldots, x0_n$ are the initial values of the geometric objects extracted from the sketch and *NR* is the function which symbolizes the Newton-Raphson numerical resolution. A triangular form is obtained with the projection functions pr_i:

$$x_1 = pr_1(NR(p_1, \ldots, p_k, x0_1, \ldots, x0_n))$$
$$\ldots$$

$$x_n = pr_n(NR(p_1, \ldots, p_k, x0_1, \ldots, x0_n))$$

These definitions are added to the construction plan. So, the numerical aspects remain internal and the Newton-Rapshon method is well integrated in our formal way of solving by triangulation.

Examples

In this section, we deal with two examples. The first example is a small one and we give all the details of the construction. The second example is a bigger one including about 70 points, 20 circles or arcs of circle, and 50 lines.

It is difficult to make a complexity analysis. We can say that the complexity for the supervisor and the algebraic agent is polynomial due to the searches in respectively the graph of subsystems and in a set of terms which does not rise. The problem comes from the knowledge based systems which are time-consuming processes. The complexity depends on the number of rules of production, which can be increased, and the problem submitted. As we did, strategies are usually chosen to improve the running time for the problems we want to solve. So, we have to make practical experiences with our own examples due to the lack of benchmark in this field. We noticed that the number of subsystems is always under 20 and the time for the problems containing less than 100 points is under 15 seconds to be solved with a Silicon Graphics R4400.

Example 1

Let us consider the constraint system of the second section. The Figure 3 shows the corresponding dimensioned sketch in YAMS. We will show how to solve this problem with the agent having the highest priority.

Each subfigure is numbered by YAMS and the corresponding geometric unknowns are suffixed by the number in dotted notation. In the subfigure 1, the agent first defines the location of a point and a line by solving an incidence constraint, for instance $onl(p7, l2)$, using the sketch data. Thus, we have:

    ** coordinates of p7.1 according to the sketch coordinates
    p7.1 = initp(-296.958, -248.927)
    ** line passing through p7.1 and having the same slope than in the sketch

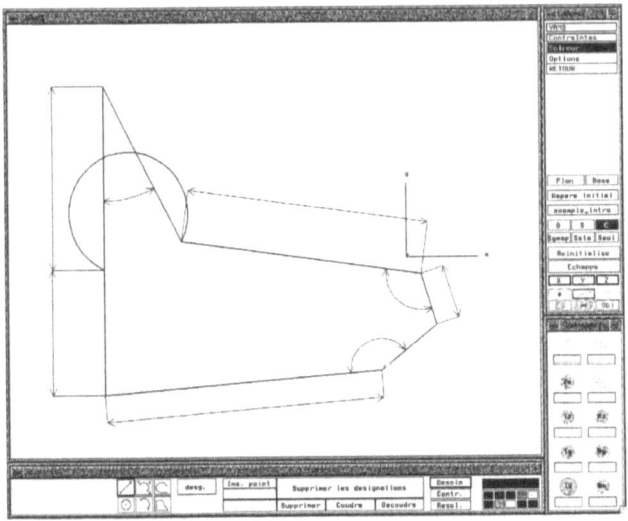

Figure 3: A screen dump of YAMS

$l2.1 = \text{initd}(p7.1, 1.56)$

These two definitions fix the location of the construction in the Euclidean plan. The construction plan grows up with definitions of elements determined by intersection between lines and circles. For instance, the point $p8.1$ is defined by the intersection of the line $l7.1$ perpendicular to $l2.1$ and the line $l8.1$ bisector of α_1. The agent can produce the subfigure 1 showed in Figure 4.

This agent is not powerful enough to continue the first construction and gives back the control to the supervisor which launches the same geometrical agent again. In the same way, the agent solves two others subfigures numbered 2 and 3 and represented in Figures 5 and 6. In the subfigure 2, the fixed elements are $p5.2$ and $l4.2$. In the subfigure 3, the fixed elements are $p7.3$ and $l6.3$.

The unknowns of the border of subfigure 1 are $p7$ and $p3$. In subfigure 2, the border unknowns are $p5$ and $p3$. Thus, we have two border constraint systems each containing two points with the known distance between them. These constraints allow construction of the point $p3$ in the subfigure 3 as shown in Figure 6.

The subfigures 2 and 3 can be assembled because points $p5$ and $p3$ are a common reference. To assemble them, the supervisor chooses to

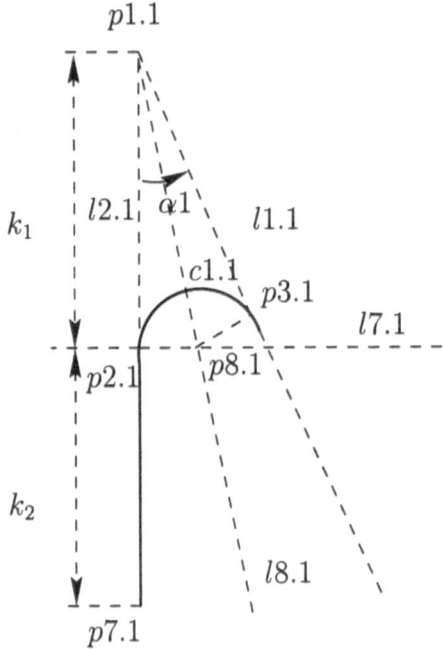

Figure 4: Subfigure 1

move subfigure 2 in order to overlap the pairs of points $(p5.2, p3.2)$ and $(p5.3, p3.3)$ (see Figure 7a). The displacement computation is represented by the definition of the symbol $dep1$. So, subfigure 2 disappears from the solving process, and subfigure 3 grows up.

The supervisor does the same with subfigures 3 and 1 and points $p7$ and $p3$ (see Figure 7b). The displacement is $dep2$. The two successive assemblings are presented in Figure 7.

So, we see that the agent cannot solve this problem on its own. Nevertheless, with the use of the invariance under displacement property and the assembling, the multi-agent system can produce a solution. The prototype YAMS finds 4 solutions and computes them in less than one second.

Example 2

The left part of the Figure 8 is the dimensioned sketch of a desk lamp. In order to keep simple the work of the designer, the different parts of

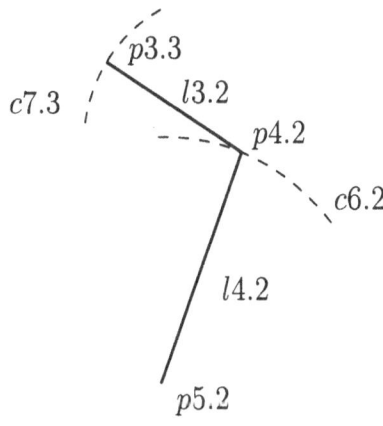

Figure 5: Subfigure 2

the lamp are constrained separately. Some constraints are then added, like equal relations between points, angles between lines, in order to put together the parts. The incidence constraints are automatically deduced from the sketch. This example requires about 150 constraints from the user. Some dimensional constraints are drawn in the figure with thin lines. The time for solving this problem is about 8 seconds.

Unfortunately, the number of solutions is too big. So, we independently solved the different parts and select for each of them the desired solution. Next the assembling constraints are added. Although the assembling process is made internally in the multi-agent system, we see that the designer may influence decomposition of the constraint system into subsystems. The solution is shown at the Figure 8.

Conclusion

In this paper, we have shown that, by using the invariance under displacements encountered in the CAD constraint systems, it is possible to break such systems into smaller ones easier to solve. This approach is mathematically described and used in a prototype which formally solves constraint systems stemming from dimensioned sketches.

The geometric universe is not limited and new types of geometric elements or constraints can be added. Moreover, this framework can in-

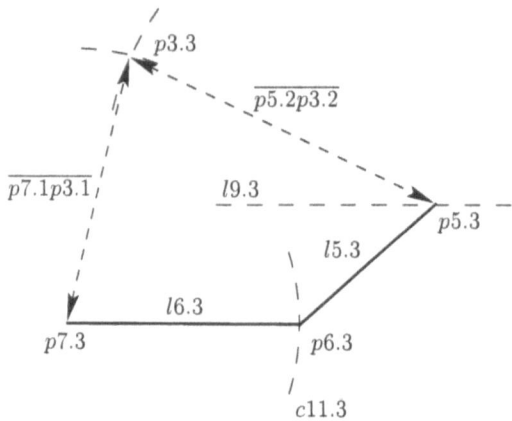

Figure 6: Subfigure 3

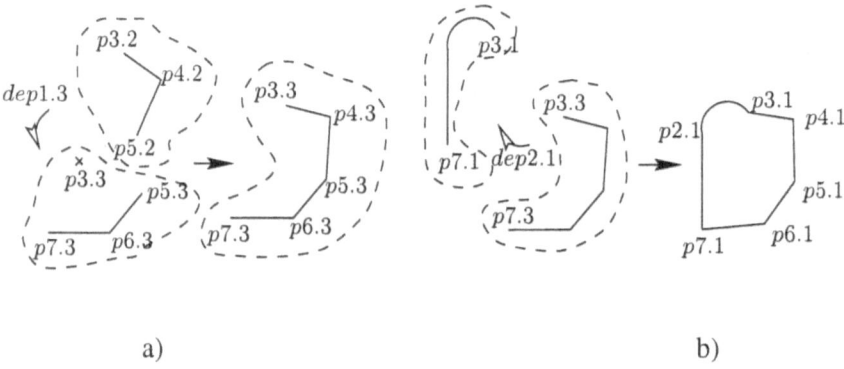

a)   b)

Figure 7: Two assemblings

tegrate any of the solving methods proposed in the literature as a local solver including those which are founded on numerical iterations or computer algebra.

In the future, our prototype should be enriched by new methods, solving strategies, heuristics for selecting methods and clever strategies to place the references. Besides, the triggering and the use of the methods can be revised in a parallel and cooperative strategy.

Our prototype needed to revisit the computer-human interface of the *Topofil* modeling tool [BDFL93, BD94] in order to visualize, enter and modify constraints. Nevertheless, the automatic recognition of implicit constraints in the way of Aldefeld would be useful [AMRV92]. Furthermore, an efficient numerical interpretation of construction plan with mul-

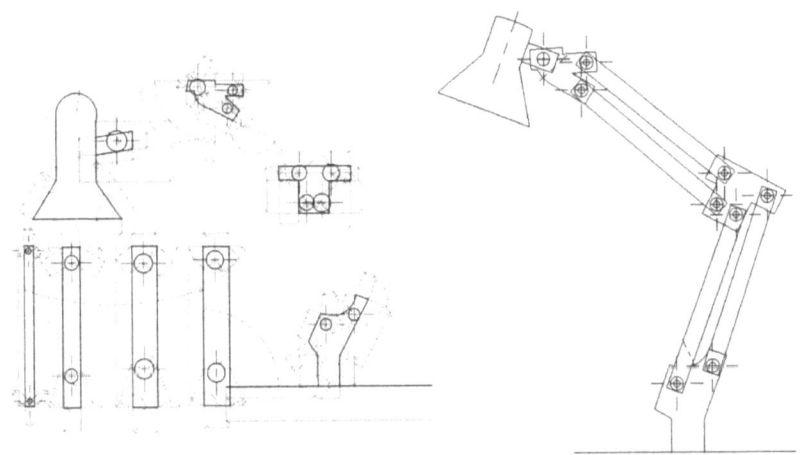

Figure 8: A constrained sketch of a desk lamp and a solution

tifunctions needs drastic pruning of the multiple solutions.

A natural extension of such a modeling tool consists in managing a rich parameterization of geometric objects. An application would then be the building of objects with the same topology, but with different forms and dimensions. Another one, as in [Kra90, Kra91], is the generation of animated sequences by a step by step progression of parameter values. This subject is studied now.

References

[Ald88]   B. Aldefeld. Variations of geometries based on a geometric-reasoning method. *Computer-Aided Design*, 20(3):117–126, 1988.

[AMRV92] B. Aldefeld, H. Malberg, H. Richter, and K. Voss. *Rule-based variational geometry in Computer-Aided Design*, pages 27–46. Artificial Intelligence in Design. (D.T. Pham ed.), Springer-Verlag, 1992.

[BD94]   Y. Bertrand and J.-F. Dufourd. Algebraic specification of a 3D-modeller based on hypermaps. *Computer Vision - GMIP*, 56(1):29–60, 1994.

[BDFL93] Y. Bertrand, J.-F. Dufourd, J. Françon, and P. Lienhardt. Algebraic specification and development in geometric modeling. In *Proceedings of the TAPSOFT Conference, LNCS 668*, pages 75–89. Springer-Verlag, 1993.

[BFH+95] W. Bouma, I. Fudos, C. Hoffmann, J. Cai, and R. Paige. A geometric constraint solver. *Computer-Aided Design*, 27(6):487–501, 1995.

[Bru93]   B. Bruederlin. Using geometric rewrite rules for solving geometric problems symbolically. *Theoritical Computer Science*, pages 291–303, 1993.

[But79]   M. Buthion. Un programme qui résout formellement des problèmes de constructions géométriques. *RAIRO Informatique*, 13(1):73–106, 1979.

[DMS97]   J.-F. Dufourd, P. Mathis, and P. Schreck. Formal resolution of geometrical constraint systems by assembling. *Proceedings of the 4th ACM Solid Modeling conf.*, pages 271–284, 1997.

[DMS98]   J.-F. Dufourd, P. Mathis, and P. Schreck. Geometric construction by assembling solved subfigures. *Artificial Intelligence Journal*, 1998.

[DS94]   J.-F. Dufourd and P. Schreck. Un système à base de connaissances pour les constructions géométriques. In *Actes de la Conférence AFCET-RFIA, Paris*, pages 351–361, 1994.

[Kra90]   G.A. Kramer. *Geometric reasoning in the kinematic analysis of mechanisms*. PhD thesis, University of Sussex, 1990.

[Kra91]   G.A. Kramer. Using degrees of freedom analysis to solve geometric constraint systems. In *Proceedings of the 1th ACM Symposium of Solid Modeling and CAD/CAM Applications*, pages 371–378. ACM Press, 1991.

[Leb50]   H. Lebesgue. *Leçons sur les constructions géométriques*. Gauthier-Villars, Paris, 1950.

[LM95]   H. Lamure and D. Michelucci. Decomposition of 2D constraints graphs. Technical report, Ecole des Mines, Saint-Etienne, 1995.

[Mat97]   P. Mathis. *Constructions gomtriques sous contraintes en modlisation base topologique*. PhD thesis, Université de Strabourg, 1997.

[Mor83]   J.L. Moris. *Computational Methods in Elementary Numerical Analysis*. John Wiley, 1983.

[Owe91]   J. Owen. Algebraic solution for geometry from dimensional constraints. In *Proceedings of the 1th ACM Symposium of Solid Modeling and CAD/CAM Applications*, pages 397–407. ACM Press, 1991.

[Pet90]   J. Petersen. *Problèmes de constructions géométriques*. J. Gabay (new edition), 1990.

[PTWF92] W.H. Press, S.A. Teukolsky, W.T Werling, and B.P. Flannery. *Numerical Recipes in C*. Cambridge University Press, 1992.

[Sch94a]   P. Schreck. Implantation d'un système à base de connaissances pour les constructions géométriques. *Revue d'Intelligence Artificielle*, 8(3):223–247, 1994.

[Sch94b]   P. Schreck. A knowledge-based for solving geometric constructions problems. In *Proceedings of the 7th International Conference on Systems research, Informatics and Cybernetics*, pages 19–24. J.W. Brahan and G.E. Lasker, 1994.

[WM95]   M. Wooldridge and J.P. Muller. Intelligent agents, theory and practice. *Knowledge Engeneering Review*, 10(2):115–152, 1995.

Qualitative Study of Geometric Constraints

H. Lamure, D. Michelucci[1]

Modeling by geometric constraints is a promising method in the field of CAD/CAM. However, this process, closely related to computer programming, is also error prone. A geometric constraints based modeler should help the end user to find his mistakes, or, better, not to commit ones by watching the building process. The well known main cause of errors with these methods is the specification of redundant constraints, and sometimes conflicting constraints. It's also important to detect under-constrained parts of the system involving indecisiveness. This chapter alludes to some numerical and probabilistic tools that could be used for this goal. There's also a decomposition method of constraints systems using tools of the graph theory. Since these two approaches have their own advantages it's worth combining them. All these methods will be studied first for systems of equations and then extended to the more specific case, but also more interesting for modeling, of systems of geometric constraints.

Introduction

In CAD-systems, *Geometric Modeling by Constraints*, also called *Variational Modeling*, enables designers to describe shapes by interactively editing a sketch and specifying geometric or engineering constraints: see R. Anderl & R. Mendgen for a survey [AM95].

Engineering constraints typically involve material properties or manufacturing parameters, and are best expressed by equations. Geometric constraints fundamentally involve points, lines, circles: typical ones specify the distance between two geometric elements (two points, a point and

---

[1]École des Mines, 158 Cours Fauriel, 42023 Saint-Étienne Cedex 02, France

a line, two parallel lines), the angle or the parallelism between two lines, the tangency between a line and a circle or between two circles, the incidence between a point and a line or a circle, etc. Actually, any algebraic equation involving coordinates is a possible constraint, as far as it is independent of the used coordinates system.

The usual geometric constraints can be represented either by algebraic equations, or by predicates. The predicates formulation has the disadvantage to restrict the set of expressible geometric constraints, and to prevent the merging of geometric constraints and of engineering ones, but it seemed more suited to rule-based approaches than the equational formulation. Up to now, all rule-based or qualitative approaches (as opposed to numerical methods) [Brü86, VSR92] rely on the predicates formulation. In contrast, this chapter will show that the equational formulation also permits a qualitative study as well.

Solving Systems of Constraints

Constraints-based modelers have to solve in some way the set of constraints by one of the following approaches: symbolic, numerical or decomposition methods.

Symbolic methods resort to tools from Computer Algebra, like resultants or Grobner Bases [AM95]. Due to their exponential running-time, they can be used only for small systems.

Numerical methods typically improve the initial and rough guess interactively provided by the user during the sketching stage, by using some Newton iterations [AM95], or the homotopy method [LM96] which has a more intuitive convergence.

Decomposition methods reduce constraint systems into basic problems, the solutions of which are then stuck back together (to quote a few: [Owe91, VSR92, BFH+95, But79]). In $2D$ basic problems are triangles: the relative location of their 3 vertices are determined by 3 constraints (*e.g.* either 3 distances or 2 distances + 1 angle or 1 distance + 2 angles); quadrilaterals or parallelograms [VSR92]; or other systems soluble by 'ruler and compass' like Appolonius's problem. The idea is that the basic problems can be solved by applying a simple mathematical formula, and the merging stage only needs some displacement. The decomposition may be performed either implicitly, by the matching process of some inference engine, firing rules [But79, VSR92, DMS97] or Prolog predicates [Brü85, Brü88], or explicitly, by searches in the graph of constraints [BFH+95, Owe91]. Up to now, this method is often restricted to $2D$ or $2D\frac{1}{2}$ applications. Moreover the graph-based approach often assumes that the graph of constraints verifies some strong assumptions, for instance being biconnected [Owe91], or hierarchically reducible into triangles [BFH+95], or that all geometric elements have exactly 2 degrees of freedom (*ie* the radius of all circles must be known), or that each constraint involves exactly 2 geometric elements. The graph-based approach proposed in this chapter overcomes these limitations. Another graph-based approach is presented in another chapter, it uses degrees of freedom analysis in Kramer's wake [Kra91].

The Need for a Qualitative Study

Specifying constraints is a rather abstract task and so a very error-prone process, like programming. To be truly user-friendly, it is crucial for geometric modelers by constraints to help their users to debug and tune their systems of constraints: It is not enough that the resolution stage detects some problem and informs the user that "there is a problem somewhere in his system of 100 equations and 100 unknowns"; the user wants a precise diagnosis like: "Among these 100 equations, these 3 are redundant and you can remove anyone of them; on the other hand these 3 unknowns are under-constrained: one equation is missing". As an aside, it is the reason why we are a little reluctant about using an optimization scheme to solve constraints: there is always a solution! and it is much more difficult for the user –and for the modeler– to find why the solution is not the expected one and where the mistake is.

Moreover some qualitative studies of systems of equations permit to decompose systems of constraints into simpler ones. Not only does it speed up the resolution, but it can also make it possible to solve a problem that would not be solved otherwise: for instance symbolic methods apply only to small enough systems. Sometimes, this decomposition can also help the user to improve his understanding of the properties of his problem.

Systems of Equations *versus* Systems of Geometric Constraints

A well behaved *system of equations* is supposed to determine n unknowns by n independent equations.

Systems of constraints met in Geometric Modeling by Constraints are special systems of equations, since well behaved systems of constraints: distances, angles, tangencies,... between geometrical elements, or engineering equations, are *independent of the used system of coordinates* (assuming it is orthonormal), and thus determine unknown coordinates of points only up to a displacement in space. For instance, in a well behaved system of constraints in $2D$, three equations are missing to completely determine unknown coordinates of points, since a $2D$ displacement is specified by three numbers: say a translation relatively to x and y axis and a rotation around the origin. In $3D$, six equations are missing, and $d(d+1)/2$ in $\mathbb{R}^d$.

To avoid ambiguities, we will use the terms : "well-constrained", "under-constrained" and "over-constrained" only for systems of equations. We will use the terms: "rigid", "under-rigid" and "over-rigid" for systems of geometric constraints.

What if one wants to specify a constraint depending on the coordinates system, say if one wants to specify the abscissa of a point A ? The equation $A_x = v$ clearly depends on the coordinates system... The solution is to explicitly represent the coordinates system Oxy by three vertices: O, X, Y, and to specify the three constraints: $||OX|| = 1$, $||OY|| = 1$ and angle$(OX, OY) = \frac{\pi}{2}$. Then $A_x = v$ can be represented by a licit geometric constraint on the (signed) distance between OY and A: dist$(OY, A) = v$.

We will impose a last restriction on geometric constraints. To solve possibly bad-constrained systems: $f_{i=1..n}(x) = 0$, some people try to minimize or to make vanish the sum of the squares of the equations: $\sum_{i=1}^{n} f_i(x)^2$ (we think it is not a so good idea, but it is not the concern here). If the $f_i = 0$ are independent on the coordinates system, so is $\sum_{i=1}^{n} f_i(x)^2 = 0$. However we will forbid the use of such a trick. Or in other words, our methods are abused by this trick.

Content of the Chapter

This chapter will propose two methods for the qualitative study of systems of equations and for systems of constraints: the first method is a probabilistic numerical one, the second one stems from graph theory and does not consider the numeric details of equations, or geometric constraints. So it can be applied to geometric constraints represented by predicates (in opposition to equations), say Prolog predicates like in Brüderlin's works [Brü86]. Of course it also applies to equations, or geometric constraints represented by equations, but only unknown-equation incidences are taken into account.

The probabilistic and numerical method is presented first. It is first applied to systems of equations and then extended to systems of geometric constraints. Then this chapter presents the graph-based approach; similarly it is first applied to systems of equations and then extended to systems of constraints; a detailed example illustrates the process. This chapter ends with possible extensions and improvements.

We only consider systems of equations or constraints, and not inequalities, because the qualitative study of such systems is too much difficult: from elimination theory, solving well-constrained systems of algebraic equations (having a finite number of roots) is simply exponential, whereas solving systems of equations and inequalities is doubly exponential [HM93]. Moreover, inequalities are the more often used only to select the wanted solution in the finite but very big solution set of a system *without* inequalities: so we can at least diagnose this last system.

The Numerical Probabilistic Method

The Qualitative Study of Systems of Equations

For the qualitative study of systems of equations to be feasible, some "genericity hypothesis" are needed. The weaker one is always satisfied in our context, where equations are typically algebraic ones, with real coefficients: exactly n independent equations are needed to determine n unknowns in $\mathbb{C}^n$. Actually, since we are interested only by real solutions and though it is a bit cavalier, we will assume a little stronger hypothesis and replace $\mathbb{C}^n$ by $\mathbb{R}^n$, in spite of some algebraic singular systems like for instance $x^2 + y^2 = 0$ which has a finite number of real solutions though there is only 1 equation for 2 unknowns. More generally, we assume that the trick of taking the sum of the squares of some equations will not be used. Some of the methods presented below need stronger genericity hypothesis, given later.

By the genericity hypothesis, a system in n unknowns and n equations: $E(X) = 0$ where $X = (x_1 \ldots x_n)$ and $E = (e_1 \ldots e_n)$, is incorrect iff the jacobian $|E'| = |(\frac{\partial e_i}{\partial x_j})|$ is identically null. Mathematically speaking, things are simple. However, the symbolic computation of the determinant of the jacobian is impracticable, even for little values of n like 10: the determinant has an exponential number of terms.

A solution is to use the probabilistic scheme [Mar71, Sch80]: the value of the jacobian is computed at some random sampling points (three or four points is enough). If each time it vanishes, then there's a probability very close to 1 that the jacobian is identically zero (see previous references to have the probability of errors). Otherwise –and obviously!– it is not identically zero.

To avoid inaccuracy problems, this test can be performed for integer points and modulo some prime integer, big enough (say about 10^6 or 10^9) to decrease the risk of unlucky reductions: there is an unlucky reduction when the value of the determinant is a non-null multiple of the used prime. For relevance, real coefficients must be represented in an exact way, for instance $\sqrt{2}$ must be represented by an auxiliary unknown α and an equation $\alpha^2 - 2 = 0$.

We can even do better and compute (always probabilistically, for some random sample points) the rank or even a base of the jacobian. Before we detail this, let us get a more intuitive insight on their meaning: we consider the unknowns $X = (x_1 \ldots x_n)$ as a function of the time, t; deriving $E(X(t)) = 0$ relatively to time, we obtain $\dot{X} E' = \dot{0}$, where $\dot{X}$ is the vector of velocities of X, compatible with E. If E is correct, *ie* if all equations e_i are independent so E' has maximal rank n, then the only solution for $\dot{X}$ is $\dot{0}$: it is impossible to move the unknowns and to keep $E(X) = 0$. More generally, the solution for $\dot{X} E' = \dot{0}$ form a vectorial space: the kernel of E' which is dual to the space spanned by the equations e_i. The rank of the vectorial space $\dot{X}$ is the number of missing determinations. The number of equations minus the rank of E' gives the number of excessive equations in E.

It is worth computing a base of $\dot{X}$ and E'. It can be done in $O(n^3)$ time by standard techniques of linear algebra. An unknown x_k is fixed by the system E iff $\dot{x}_k$ is zero for all the vectors in the base of $\dot{X}$; otherwise the number of vectors with a non zero component $\dot{x}_k$ in the base for $\dot{X}$ gives the number of remaining degrees of freedom for x_k. Of course, the numerical values of vectors, for bases of $\dot{X}$ and E', depend on the random sampling points, but the structure of E' and $\dot{X}$ (rank, corank, fixed and not fixed unknowns) does not.

As an example, let us consider this little system:

$$E = \begin{cases} x^2 + y^2 - 1 = 0 \\ y^2 + z^2 - 2 = 0 \\ x^2 - z^2 + 1 = 0 \\ w^2 + 2w + 1 = 0 \end{cases}$$

whose jacobian E' is computed at random point: $p = (1\ 2\ 3\ 4)$:

$$E' = \begin{pmatrix} 2x & 0 & 2x & 0 \\ 2y & 2y & 0 & 0 \\ 0 & 2z & -2z & 0 \\ 0 & 0 & 0 & 2w+2 \end{pmatrix} = \begin{pmatrix} 2 & 0 & 2 & 0 \\ 4 & 4 & 0 & 0 \\ 0 & 6 & -6 & 0 \\ 0 & 0 & 0 & 10 \end{pmatrix}$$

$E'(p)$ has rank 3 and a base of its kernel is: $(\dot{x}\ \dot{y}\ \dot{z}\ \dot{w}) = (6\ -3\ 2\ 0)$. Thus x, y and z are not determined, but w is, since $\dot{w} = 0$ for all vectors in the kernel base.

Last but not least, this approach (probabilistically computing ranks or bases for $\dot{X}$ and E') can also be used when the number of equations is not equal to the number of unknowns. Thus it can also be performed on-line: in an interactive context, equations are likely introduced once at a time. The following interactive protocol may be convenient: the software maintains an independent set of equations E: "independent" means: rank(E')=cardinality(E) where the rank is computed with the probabilistic method. When the user proposes a new equation e, the software computes if $E' \cup \{e'\}$ is dependent or not: for instance, e' is decomposed into $e' = e_d + e_i$ by Gram-Schmidt's orthogonalization procedure, where e_d is the projection of e' in the range of E' and e_i is the orthogonal part. Again, these computations can be performed modulo some big prime (in particular, no square root is needed). If $\{e'\} \cup E'$ is independent ($e_i \neq 0$), e is inserted in E; otherwise the software asks the user which equation to remove in the minimal dependent set in $E \cup \{e\}$: to find the latter, just remove from E' each element such that remaining ones and e' are still dependent. The set of compatible velocities $\dot{X}$ is also maintainable on-line.

The probabilistic numerical method is simple and very easy to implement. It may use the sparseness of the system at hand in order to speed up computations. Anyway it cannot be slower than the numerical resolution method. It can distinguish between dependent and independent subsets of equations on one part, and between fixed and unfixed unknowns on the other part.

The Qualitative Study of Systems of Geometric Constraints

From Equations to Geometric Constraints

Since the well behaved systems of constraints are under-constrained systems of equations (as already explained), it may seem this method does not directly apply to systems of geometric constraints.

A first straightforward way to overcome this limitation is to add some ad-hoc equations. If for instance in $2D$, specify for some couple of points A and B that $A_x = A_y = 0$ and $B_y = 0$. In $3D$ specify for some triplet of points A, B and C that $A_x = A_y = A_z = 0$, $B_y = B_z = 0$ and $C_z = 0$. Then all the numerical probabilistic approach directly applies: it is possible to know the fixed and not fixed unknowns, and to detect dependent subsets of equations.

A second way is not to add ad-hoc equations, but to remember that a rigid system must have exactly 3 (respectively 6) remaining degrees of freedom in $2D$ (respectively $3D$), *ie* the vectorial space of compatible velocities $\dot{X}$ must have rank 3 (respectively 6). In $\mathbb{R}^d$, it must have rank $d(d+1)/2$.

An Example

Let us study the $2D$ system in Fig. 1 (the same example is also studied with the graph-based method in Fig. 7). Edges represent constraints of distance between points. It is intuitively obvious that, though there is the good number of constraints (6 points, $2 \times 6 - 3 = 9$ constraints) for the graph to be rigid, the subset $P_1 P_2 P_5 P_4$ is over-rigid and the subset $P_2 P_3 P_6 P_5$ is under-rigid.

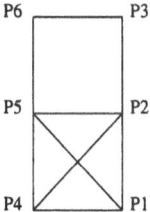

Figure 1: A simple $2D$ system of geometric constraints. Each edge represents a constraint of distance between its two vertices.

Distance constraints are in this order:

$$E = (P_1 P_2 \quad P_1 P_4 \quad P_2 P_4 \quad P_2 P_5 \quad P_4 P_5 \quad P_1 P_5 \quad P_2 P_3 \quad P_3 P_6 \quad P_5 P_6)$$

The vector of unknowns is $X = (x_1\ y_1\ x_2\ y_2 \ldots x_6\ y_6)$. Thus the jacobian is (after division by 2) $E' =$

$$\begin{pmatrix}
x_1-x_2 & x_1-x_4 & 0 & 0 & 0 & x_1-x_5 & 0 & 0 & 0 \\
y_1-y_2 & y_1-y_4 & 0 & 0 & 0 & y_1-y_5 & 0 & 0 & 0 \\
x_2-x_1 & 0 & x_2-x_4 & x_2-x_5 & 0 & 0 & x_2-x_3 & 0 & 0 \\
y_2-y_1 & 0 & y_2-y_4 & y_2-y_5 & 0 & 0 & y_2-y_3 & 0 & 0 \\
0 & 0 & 0 & 0 & 0 & 0 & x_3-x_2 & x_3-x_6 & 0 \\
0 & 0 & 0 & 0 & 0 & 0 & y_3-y_2 & y_3-y_6 & 0 \\
0 & x_4-x_1 & x_4-x_2 & 0 & x_4-x_5 & 0 & 0 & 0 & 0 \\
0 & y_4-y_1 & y_4-y_2 & 0 & y_4-y_5 & 0 & 0 & 0 & 0 \\
0 & 0 & 0 & x_5-x_2 & x_5-x_4 & x_5-x_1 & 0 & 0 & x_5-x_6 \\
0 & 0 & 0 & y_5-y_2 & y_5-y_4 & y_5-y_1 & 0 & 0 & y_5-y_6 \\
0 & 0 & 0 & 0 & 0 & 0 & 0 & x_6-x_3 & x_6-x_5 \\
0 & 0 & 0 & 0 & 0 & 0 & 0 & y_6-y_3 & y_6-y_5
\end{pmatrix}$$

A possible base for compatible velocities $\dot{X}$ is:
- an x-translation: $(1\ 0\ 1\ 0\ 1\ 0\ 1\ 0\ 1\ 0\ 1\ 0)$
- an y-translation: $(0\ 1\ 0\ 1\ 0\ 1\ 0\ 1\ 0\ 1\ 0\ 1)$
- a rotation around the origin, for instance: $(-y_1\ x_1 -y_2\ x_2 \ldots -y_6\ x_6)$

- the last vector is for the possible motion of P_3 and P_6 around P_2 and P_5 respectively. Any non-null vector $(0\ 0\ 0\ 0\ \dot{x}_3\ \dot{y}_3\ 0\ 0\ 0\ 0\ \dot{x}_6\ \dot{y}_6)$ such that

$$(\dot{x}_3\ \dot{y}_3\ \dot{x}_6\ \dot{y}_6) \begin{pmatrix} x_3 - x_2 & x_3 - x_6 & 0 \\ y_3 - y_2 & y_3 - y_6 & 0 \\ 0 & x_6 - x_3 & x_6 - x_5 \\ 0 & y_6 - y_3 & y_6 - y_5 \end{pmatrix} = (0\ 0\ 0)$$

is suitable.

For short, we don't give symbolic values for this last vector, nor details of probabilistic computations. Anyway, the computed rank of $\dot{X}$ is four: three are due to normal $2D$ rigid body motions, the fourth is due to the remaining degree of freedom in part $P_2 P_3 P_6 P_5$. Similarly, the computed rank of E' is 8, whereas E has 9 equations: redundancy is detected. Adding three ad-hoc equations like $x_1 = y_1 = x_2 = 0$, and computing $\dot{X}$ will show that P_1, P_2, P_4, P_5 cannot move, but P_3 and P_6 have still one degree of freedom.

Remark: systems like in this example, where all constraints specify distances between points, form a special class of problems, which is called the rigidity graph problem, and which has been studied by graph-theorists or combinatorists, and by structural engineers who want to guarantee the stability of frameworks in buildings. These communities call the jacobian the "rigidity matrix". They proved the correctness of the probabilistic method for the rigidity graph problem, in all dimension, just assuming weak genericity hypothesis : it is Gluck's theorem. They also found a graph-characterization for $2D$ rigid graphs : it is Laman's theorem. They proposed graph-based or matroid-based methods for testing rigidity in $2D$. Up to now, a graph-characterization for rigidity is still unknown in R^3 and beyond. See [Hen92a, LP86, Rec86].

Ergonomic Issues

The ergonomic issues are not our concern here, but in passing we propose the following visual protocol for helping the user to see which part of his sketch is still under-determined. Let X_1 be the current solution of the set of constraints $E(X) = 0$. If some part of the sketch is under-constrained, then there is some velocity vector $\dot{X}_1$ such that $\dot{X}_1 E'(X_1) = 0$. The idea is to smoothly deform the sketch from its state X_1 to a state X_2, X_3, etc but keeping $E(X_i) = 0$ for all i. X_2 is computed from $X_1 + \epsilon \dot{X}_1$, with some correction by some Newton-Raphson's iterations or some gradient descent. Idem for X_3 from X_2, and so on. It is also easy to allow the user to specify the 2 motionless points A and B in $2D$: so he will see which part remains motionless relatively to AB, *ie* the maximal rigid part containing

AB. The same holds in $3D$ for 3 motionless points. Of course, the modeler can also automatically deduce the maximal rigid parts, in the same way.

Limitations of the Numerical Probabilistic Method

The numerical probabilistic approach has limitations:

1. It cannot distinguish between fixed but well-constrained unknowns, and fixed but over-constrained ones; this notion is relevant only for sparse systems: for instance in the system $x+y+z = 2x+2y+2z = 0$, unknowns x, y and z are in the same time over-constrained (since they are involved by redundant equations) and under-constrained (since compatible velocities $\dot{x}$, $\dot{y}$ and $\dot{z}$ are not zero). But the graph-based method (next section) will solve this problem. The next two limitations are actually a reformulation of the genericity hypothesis:

2. Another limitation of the probabilistic numerical method, and *a fortiori* of the graph-based one, and in fact of all polynomial-time methods is that "subtle dependencies" (defined below) between equations cannot be detected. Assuming all equations to be polynomials, these methods cannot detect that one of the polynomials is in the ideal or in the radical of the others, or, in more intuitive words, that one equation is a consequence of the others. Mathematically, f_0 is in the ideal generated by $f_{i=1..n}$ iff there exist polynomials h_i such that $f_0 = h_1 f_1 + h_2 f_2 + \ldots h_n f_n$, and in its radical iff there exist some integer $k \geq 1$ such that f_0^k is in the ideal generated by $f_{i=1..n}$. In such cases, clearly, $f_1 = f_2 = \ldots f_n = 0$ implies $f_0 = 0$. Moreover, the associated jacobian of $f_0 = f_1 = \ldots f_n = 0$ has non-maximal rank at the common roots of $f_0 = f_1 = \ldots f_n = 0$, but it can have maximal rank elsewhere. A simple example is: $f_1 = x^2$, $f_2 = y + 1$, $f_0 = z f_1 + y f_2$; though the jacobian $\left| \frac{\partial f}{\partial x} \right| = -2x^3$ vanishes when $f_1 = f_2 = 0$, it is non identically zero.

3. In the same way, polynomial-time methods cannot detect that one of the polynomial equation, f_0, contradicts the others, *ie* that there exist polynomials h_i, an integer $k \geq 1$ and a non-null constant c such that $f_0^k = c + h_1 f_1 + h_2 f_2 + \ldots h_n f_n$. In such cases, clearly, $f_1 = f_2 = \ldots f_n = 0 \Rightarrow f_0 = c \neq 0$. Here again, the associated jacobian of $f_0 = f_1 = \ldots f_n = 0$ has non-maximal rank at the common roots of $f_1 = f_2 = \ldots f_n = 0$, but it can have elsewhere.

For example, consider the system with 4 points O, A, B, C and 5 constraints: O is the middle of AB, distance OC equals distance OA, AC is orthogonal to CB (this constraint is a consequence of the previous ones), and distance AB is given. This system is not rigid since C can freely rotate around circle with diameter AB. Equations are:

$$\begin{cases} e_1 = 2x_O - x_A - x_B = 0 \\ e_2 = 2y_O - y_A - y_B = 0 \\ e_3 = (x_C - x_O)^2 + (y_C - y_O)^2 - (x_A - x_O)^2 - (y_A - y_O)^2 = 0 \\ e_4 = (x_C - x_A)(x_C - x_B) + (y_C - y_A)(y_C - y_B) = 0 \\ e_5 = (x_A - x_B)^2 + (y_A - y_B)^2 - d^2_{AB} = 0 \end{cases}$$

Unknowns are: $(x_O\ y_O\ x_A\ y_A\ x_B\ y_B\ x_C\ y_C)$. The jacobian is:

$$E' = \begin{pmatrix} 2 & 0 & 2x_A - 2x_C & 0 & 0 \\ 0 & 2 & 2y_A - 2y_C & 0 & 0 \\ -1 & 0 & 2x_O - 2x_A & x_B - x_C & 2x_A - 2x_B \\ 0 & -1 & 2y_O - 2y_A & y_B - y_C & 2y_A - 2y_B \\ -1 & 0 & 0 & x_A - x_C & 2x_B - 2x_A \\ 0 & -1 & 0 & y_A - y_C & 2y_B - 2y_A \\ 0 & 0 & 2x_C - 2x_O & 2x_C - x_A - x_B & 0 \\ 0 & 0 & 2y_C - 2y_O & 2y_C - y_A - y_B & 0 \end{pmatrix}$$

Term $\frac{\partial e_4}{\partial x_C} = 2x_C - x_A - x_B$ is equal to $2x_C - 2x_O = \frac{\partial e_3}{\partial x_C}$, due to equation e_1. Similarly, $\frac{\partial e_4}{\partial y_C} = \frac{\partial e_3}{\partial y_C}$ due to equation e_2. These equalities don't hold for generic (*ie* random) values of x_O, x_A, x_B, x_C, $y_O \ldots y_C$ thus the compatible velocity $\dot{X} = (0\ 0\ 0\ 0\ 0\ 0\quad y_O - y_C\quad x_C - x_O)$ is missed by the probabilistic method, which wrongly finds this system is rigid.

Subtle dependencies give us a trick to change all geometric theorems into non-generic constraints systems which mislead the numerical probabilistic method. Just note that, when you don't know the trick, such systems are unlikely to occur, and the probabilistic method is still interesting. Moreover no method in polynomial time can detect subtle dependencies: one has to resort to some symbolic computation machinery, like Grobner bases, which are exponential in time, and sometimes doubly exponential. As a consequence, such symbolic methods can be used only for very small systems or subsystems (less than 10 non linear algebraic equations); the graph-based method in the next section just gives a fast way to find smallest subsystems in large sparse systems of equations.

The Bipartite Graph-Based Method

An alternative approach, stemming from graph theory, exploits the sparseness of systems of equations or geometric constraints. The idea is to consider the bipartite graph associated to a given system of equations or geometric constraints: each unknown is associated to an unknown-vertex, each equation is associated to an equation-vertex, and an edge join an unknown-vertex and an equation-vertex iff the corresponding unknown appears in the corresponding equation (Fig. 2). This graph is bipartite because there is no edge between any two unknown-vertices, and between any two equation-vertices. In the following we will be drawing equation-vertices above the unknown-vertices.

We first apply this method to systems of equations, then extend it to systems of geometric constraints.

Figure 2: *A* is well-constrained in generic case : it has a perfect matching. *B* is always singular : it has no perfect matching.

The Qualitative Study of Systems of Equations

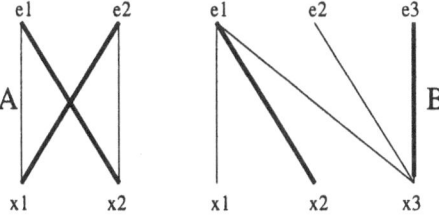

$$\begin{vmatrix} a & b & 0 \\ c & d & 0 \\ 0 & e & f \end{vmatrix} = \text{adf-bcf}$$

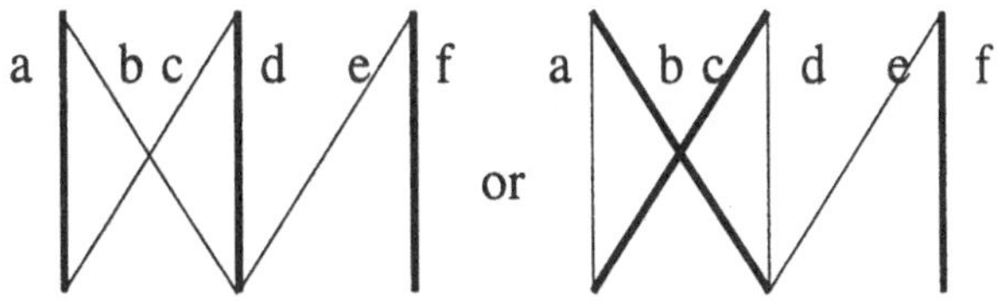

Figure 3: A graph and its perfect matchings. Note the correspondence between each perfect matching and each term of the determinant.

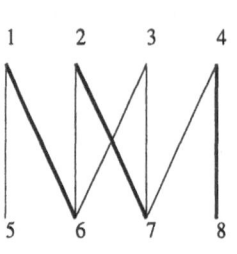 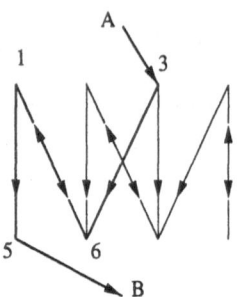 

Figure 4: To find a maximum matching, start with an initial matching (possibly empty), and improve its cardinality while possible, as follows. Built the corresponding directed graph in the middle: edges in the matching are oriented in both directions, while others are oriented downward. Then find a shortest path (here: A, 3, 6, 1, 5, B) from a non-saturated Above vertex to a non-saturated Below one. Finally invert all status (*ie* belonging or not to the matching) of edges (here 36, 61, 15) along the found path.

Dulmage-Mendelsohn's Decomposition

In the generic case (we will come back later to this point), the study of this bipartite graph first permits to decompose the system into its well-, over- and under-constrained parts [LP86]. This decomposition is due to Dulmage and Mendelsohn. Though it was not initially developed with the study of systems of equations in mind, it appears to be relevant for this problem. We will need the following definitions:

A *matching* of a graph is a subset of its edges, such that any two distinct edges of the matching never have a common vertex. A matching is *maximum* iff it is maximal in cardinality. A vertex is *saturated* by a matching iff it is a vertex of one edge of this matching. A matching saturating all vertices of a graph is called *perfect*: see Fig. 3. From an algorithmic point of view, polynomial algorithms to compute a maximum matching for bipartite graphs are known, for instance Hopcroft and Karp's method [HK73, AHU83], see Fig. 4.

The Dulmage-Mendelsohn's decomposition is illustrated Fig. 5. The main properties are, in a jumble: there is no edge between D_1 and C_2, between D_2 and C_1 and between D_1 and D_2. $G_1 = C_1 \cup C_2$ has a perfect matching, so $|C_1| = |C_2|$. D_1 and D_2 are respectively the set of equation- and unknown-vertices which are not saturated by at least a maximum matching. A_2 (respectively A_1) is the set of the neighbors of D_1 (respectively D_2). Vertices of A_2 and A_1 are all saturated by all maximum matchings. Edges between C_1 and A_2, between C_2 and A_1, between A_1 and A_2, actually edges between distint G_i, never belong to a maximum matching.

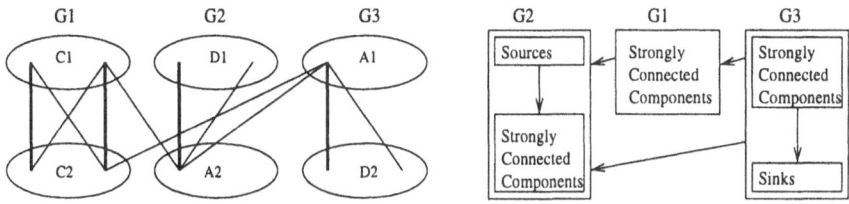

Figure 5. The Dulmage-Mendelsohn's decomposition of a graph G, with a maximum matching, and the structure of the induced directed graph G'.

The important thing for us is that $G_1 = C_1 \cup C_2$ is the well-constrained part of the system, $G_2 = D_1 \cup A_2$ its over-constrained part, $G_3 = A_1 \cup D_2$ its under-constrained part. Only the G_1 part has a perfect matching (Fig. 3).

To give a fast method for computing the Dulmage-Mendelsohn's decomposition, we need the following definitions: a *directed* graph is said to be *strongly connected* iff for any pair x and y of vertices there exists a *directed path* from x to y and from y to x. The *strongly connected components* of a graph are its maximal strongly connected subgraphs, they partition its vertices. *Strongly* connected components must not be confused with connected components: the latter does not take into account the orientation of edges.

Let M be any maximum matching of the bipartite graph G. G' is the *directed* graph obtained from G by replacing each edge (x, y) in M by two arcs xy and yx, and by orienting all other edges from equation-vertices to unknown vertices. The *strongly connected components* of G' are included either in G_1, or in G_2, or in G_3. Moreover if there are non saturated equation-vertices, then they are the sources of G_2. Symmetrically, if there are non saturated unknown-vertices, then they are sinks of G_3. Thus G' has the structure shown in Fig. 5. An algorithm to compute the Dulmage-Mendelsohn's decomposition follows: Find a maximum matching M of G; build the directed graph G' from G; G_2 is the set of all descendants of sources of G'; symmetrically G_3 is the set of all ancestors of sinks of G'; finally $G_1 = G' - G_2 - G_3$. Of course, this decomposition is unique and does not depend on the initial maximum matching. Finding a maximum matching can be done in $O(e\sqrt{v})$, where e is the number of edges of G and v its number of vertices, by using Hopcroft and Karp's method [HK73, AHU83]. The other steps can be done in $O(e + v)$: the computation of ancestors and descendants is made by a classical depth first or breadth first search in linear time. Clearly $e = O(v^2)$. In practice, e is the more often proportional to v. Anyway, we have a fast method to diagnose a system of equations.

Finding the Irreducible Subsystems

Secondly, for well-constrained systems, the study of the associated bipartite graph permits to find its irreducible subsystems and the dependencies between them. This reduction greatly speeds up the resolution process, whatever the resolution method used, numerical or symbolic. Moreover, if the use of this decomposition is visible for the user, it allows him to follow the resolution process step by step, *ie* irreducible after irreducible: the decomposition gives the running trace of the resolution process and makes it more self explanatory.

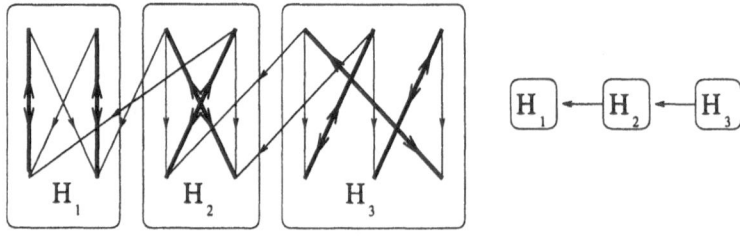

Figure 6: Graph G', and its reduced graph.

Let G the graph of a well-constrained system, thus $G = G_1$, $G_2 = G_3 = \emptyset$ and G has a perfect matching (in passing, it is König-Hall's theorem). An irreducible subgraph of G is a minimal subset of k equation-vertices and *all* their k adjacent unknown-vertices: it corresponds to a minimal subsystem of k equations involving k unknowns, which can be solved independently. Fig. 6 shows an example of a well-constrained graph G with an irreducible subgraph H_1; $G - H_1$ has an irreducible subgraph H_2; The remaining graph $G - H_1 - H_2$ is itself irreducible. On this example (see Fig. 6), one can see that H_1, H_2 and $H_3 = G - H_1 - H_2$ are just the *strongly* connected components of the oriented graph G', obtained from G in the same way than in the previous section. Actually, it is a general property [LP86], which gives us a fast method to compute the irreducible subsystems of a given system: a perfect matching M of G is first computed; the directed graph G' is built as before; the *strongly* connected components of G' are computed, say by Tarjan's linear-time method [Tar72, AHU83]: they are the searched subsystems, and they do not depend on the initial perfect matching. To obtain the dependencies between the irreducible subsystems, build the reduced graph R from G' by contracting each *strongly* connected component in a vertex (Fig. 6): each remaining arc in R, say from s_1 to s_2, means that subsystem s_1 uses some unknowns of s_2, thus s_2 has to be solved before s_1. Of course R is acyclic and defines a partial order between the subsystems. A compatible total order between subsystems can be obtained by any topological sorting of R, or as a byproduct of Tarjan's method: it is the order in which the *strongly* connected components are obtained. All steps are linear, except the determination of a starting matching, in $O(e\sqrt{v})$. In practice, the time for the decomposition is always negligible in front of the resolution time. For reducible systems, speed up factors of 10 or more are usual [AAJM93].

Genericity Hypothesis

The graph-based approach uses only the structure of the system and forgets all numerical informations. Thus it assumes stronger genericity hypothesis than the numerical probabilistic approach. For instance the graph A in Fig. 2 is associated to singular systems, in degenerate cases (like: $x + y = 2$, $2x + 2y = 4$ which has an infinity of solutions or like $x + y = 2$, $2x + 2y = 3$ which has no solution at all), and also to well-constrained systems, in the generic case. The probabilistic approach can discriminate (up to limitations, as already seen) between the non-generic and generic cases when the graph-based approach obviously cannot. Note that the nullity of the jacobian in degenerate cases has nothing to do with the structure of the system, and that it suffices to randomly and infinitesimally perturb the coefficients of such degenerate systems (without changing their structure) to recover the generic situation; or, in other words, degenerate cases occur with probability 0 [LP86]. On the other hand, when the graph-based approach states that a system is not well-constrained, then it is not (see graph B in Fig. 2) and no matter the values of the coefficients are. Mathematically speaking, a sufficient condition for a system to be *generic* is that its coefficients are algebraically independent (of course it is a very unrealistic assumption, since coefficients are integers or floating point numbers, *ie* rational numbers ...) so that they do not verify any parasitic condition, and the rank of the jacobian is then strictly equal to the cardinality of the maximum matching of the corresponding bipartite graph. Degeneracies can only decrease the rank.

The Qualitative Study of Systems of Geometric Constraints

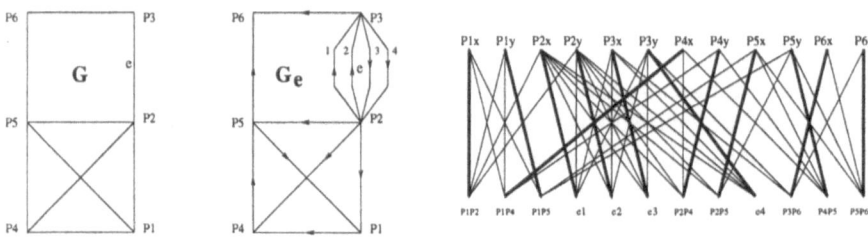

Figure 7: This graph G is incorrect, *ie* non rigid. However, adding some three ad-hoc equations (the quadrupled edge: e) gives a graph G_e, whose associated bipartite graph $Bip(G_e)$ is well-constrained. The perfect matching of $Bip(G_e)$ is drawn directly on G_e by orienting its edges.

The extension from system of equations to systems of geometric constraints is not simple for the bipartite graph-based approach. There are cases where adding some ad-hoc equations produces a bipartite graph having a perfect matching, *ie* indicating that the system is rigid, though it is not. Such a case is illustrated in $2D$ in Fig. 7: the vertices of this graph of constraints represent, say, points with unknown vertices, and edges represent constraints of distance. In Fig. 7, one of the original constraints has been quadrupled: we pin the corresponding edge e on the plane. Call G_e the resulting graph. The bipartite graph $Bip(G_e)$ has a perfect matching, though the initial system of constraints is obviously incorrect: the square with its two diagonals is over-rigid, the other square is under-rigid.

The perfect matching has been displayed in a very compact way, directly on the graph as an orientation of its edges: an edge (a, b) in a graph of constraints G_e is oriented from a to b iff, in $Bip(G_e)$, the edge (ab, a_x) or (ab, a_y) belongs to the matching. The fact that the matching is perfect implies that each vertex in G_e receives exactly 2 arcs.

However, pinning another edge of the graph in Fig. 7 permits to detect that this graph is over-rigid in one part and under-rigid in the other part. We just have to pin each and every of its edge. We now see a general formulation of the method.

Hypothesis

For the sake of simplicity, we assume that the system of constraints involves n unknown $2D$ points, the coordinates (x_i, y_i) of which depend on the coordinates system, and k unknowns $u_1 \ldots u_k$ which are independent of the coordinates system: they can be (possibly signed) distances, (possibly signed) areas, scalar products, angles, radii of circles... or non geometric unknowns.

Moreover, all constraints must be independent of the coordinates system: if (x_i, y_i, u_j) is a solution, so is $(x_i + T_x, y_i, u_j)$ for any value T_x (a translation in x has been applied), so is $(x_i, y_i + T_y, u_j)$ for any value T_y (a translation in y has been applied), and so is $(x_i \cos \theta - y_i \sin \theta, x_i \sin \theta + y_i \cos \theta, u_j)$ for any value of θ (a rotation around the origin has been applied). Such a constraint either involves no points, or involves at least 2 points, but never a single point. Note that we accept constraints involving more than two points: these systems are more general than the graph of constraints illustrated by Fig. 7 or 8, where constraints must involve exactly 2 geometric elements (since an edge has 2 vertices). According to this criteria the bipartite graph-based approach has a larger expressive power than the one relying on graphs of constraints [Owe91, BFH+95].

Such a system of constraints has $P(n, k) = 2n + k - 3$ degrees of freedom when $n \geq 2$ and $P(n, k) = k$ degrees when $n = 0$. To fix without redundancy the $2n + k$ unknowns up to a displacement in the plane, in other words to be rigid, the system must have exactly $P(n, k)$ equations (which is trivial to check) and all its subsystems with n' points and k' other unknowns must have at most $P(n', k')$ equations: if one has more than $P(n', k')$ equations, it is over-rigid. For short, we will say that a rigid system must have the P property.

We assume that the system at hand has $P(n, k)$ equations, and that its Dulmage-Mendelsohn's decomposition has an empty part G_3 (the over-constrained part): in other words, all equations are covered by a maximal matching. Now, by slightly extending a method due to Hendrickson [Hen92b], it is possible to verify that the system fulfills P, *ie* that there is no over-determined points:

Decomposition Algorithm

For each constraint e involving at least one point (and so, at least two points as we have just seen) in the system G, consider the bipartite graph $Bip(G_e)$ where G_e is obtained from G by adding three ad-hoc equations involving the same unknowns than e: intuitively speaking, we "pin" two points appearing in e on the plane. Verify that $Bip(G_e)$ has a perfect matching: if not, the system G_e is incorrect and the Dulmage-

Mendelsohn's decomposition gives an over-constrained part in G_e, and so an over-rigid part in the initial system of constraints G. Hendrickson has proved [Hen92b] that the system G fulfills P iff all the bipartite graphs G_e have a perfect matching: the proof is not difficult but a bit lengthy and must be omitted for conciseness.

Only the first maximum matching has to be computed from scratch: the others can be obtained more quickly, in linear time, by updating the previous one (see Fig. 4). For this reason, this method may be used on-line, when constraints are added or removed one at a time, but this point is not detailed due to lack of space. Assuming that $k = O(n)$, and that each constraint involves $O(1)$ unknowns (which is generally the case), then the first step could be done in $O(n^2)$, and each of the n other steps could be done in $O(n)$, so this method will work in $O(n^2)$, whereas the probabilistic numerical method is in $O(n^3)$.

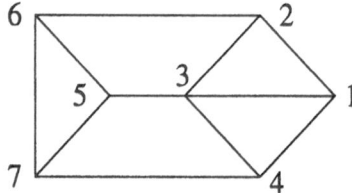

Figure 8: A graph of constraints.

Moreover, at each step (*ie* for each G_e), it is possible to compute the decomposition into irreducible parts: each irreducible part – it contains the particularized constraint – gives a rigid subsystem, assuming the system is correct. For instance, for the system of constraints in Fig. 8, when pinning edge-constraint $(1, 3)$, subsystems $\{1, 3\}$, $\{1, 2, 3\}$ and $\{1, 3, 4\}$ are found. When pinning edge-constraint $(5, 6)$ or $(6, 7)$ or $(5, 7)$, subsystem $\{5, 6, 7\}$ is found (and the trivial subsystem equal to the pinned edge itself, of course). Thus *all* rigid subsystems are found this way. If the subrigid parts are "small" (*ie* involve only a number of other subrigids independent of n) then we can use this decomposition in order to speed up the resolution process. Indeed it's possible to solve each of the small systems in constant time: by applying a resolution scheme in simple cases (for example when only 3 equations are involved), by rewriting equations, or by formal resolution [DMS97].

We have used the obvious fact that a rigid system must have the P property; when the constraints are solely distances between vertices, G.

Laman [Lam70] has been able to prove the converse: a system verifying the P property is rigid, so the P property is a full characterization of these rigid graphs. But as far as we know, this converse has not been extended up to now for any kind of $2D$ constraints.

As already seen, and for the same reasons, it is worth combining the graph-based approach and the numerical probabilistic one.

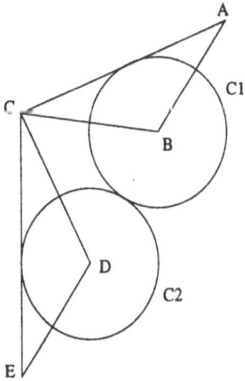

Figure 9: A $2D$ system of geometric constraints.

An Example

This section presents a simple and complete $2D$ example: see Fig. 9. The constraints are:

- $\alpha^2 - 2 = 0$ which yields $e_0(\alpha) = 0$

- $||AC|| = 4\alpha$ which yields $e_1(A, C, \alpha) = 0$

- $||AB|| = 3\alpha$ which yields $e_2(A, B, \alpha) = 0$

- $||BC|| = 2\alpha$ which yields $e_3(B, C, \alpha) = 0$

- the circle C_1 with center B and radius R_1 is tangent to the line (AC) which yields $e_4(A, B, C, R_1) = 0$

- $||CD|| = 4$ which yields $e_5(D, C) = 0$

- $||CE|| = 3$ which yields $e_6(E, C) = 0$

- $||DE|| = 2$ which yields $e_7(D, E) = 0$

- the circle C_2 with center D and radius R_2 is tangent to the line (CE) which yields $e_8(C, D, E, R_2) = 0$

- Circles C_1 and C_2 are tangent each other: $e_9(B, D, R_1, R_2) = 0$

The method detects that the system is rigid and its rigid subsystems are: $S_0 = \{e_0\}$ which determines α, $S_1 = S_0 \cup \{e_1, e_2, e_3\}$ which determines the triangle (A, B, C), $S_2 = \{e_5, e_6, e_7\}$ which determines the triangle (C, D, E), $S_3 = S_1 \cup \{e_4\}$ which determines the radius R_1, $S_4 = S_2 \cup \{e_8\}$ which determines the radius R_2, $S = S_3 \cup S_4 \cup \{e_9\}$ which permits to assemble the 2 parts of the figure. For simplicity we have omitted the trivial subsystems $\{e_1\}, \{e_2\}, \{e_3\}, \{e_5\}, \{e_6\}, \{e_7\}$.

Possible Extensions

Combining Numerical and Graph Methods

It is worth combining the two approaches. The graph-based approach gives a decomposition into over-, under- or well-constrained parts, or into irreducible well-constrained parts, which the probabilistic and numerical method cannot produce. However these parts cannot be studied further by the graph-based approach.

Inside each irreducible and well-constrained part, the numerical probabilistic method may then be used, to detect redundancies between equations or not fixed unknown(s) (probably caused by some mistakes from the user) that could not be detected by the graph-based approach. Recall the graph-based approach can be abused by non generic systems which will not abuse the probabilistic and numerical method which uses a weaker genericity hypothesis.

The fact that the graph-based approach is more easily abused than the probabilistic one must not be thought as a defect of the former: on the contrary, the fact that a system is found well-constrained for the former approach, and bad-constrained for the latter one is very informative, when debugging a set of constraints.

The $3D$ Case

We have not investigated the use of bipartite graphs for $3D$ systems of constraints so far. The function $P(n, k)$ becomes in $3D$: $P(0, k) = k$,

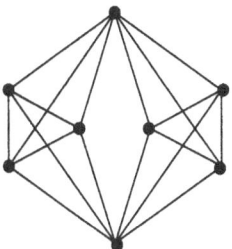

Figure 10: Here is the classical counterexample of G. Laman's proposition in $3D$. This graph fulfills the P property, but it is not rigid. On one hand, there is no reason for the height of the left half to be equal to the height of the right half, and on the other hand, the two halves can freely rotate around the vertical axis of symmetry.

$P(2, k) = 1 + k$ and $P(n, k) = 3n - 6 + k$ when $n \geq 3$. A system with n $3D$ points and k other unknowns must have exactly $P(n, k)$ equations (which is trivial to check) and all its subsystems with n' points and k' other unknowns must have at most $P(n', k')$ equations: if one has more than $P(n', k')$ equations, it is over-determined. Hendrickson's method has to be modified in this way: for each couple of constraints involving at least a common point, we must add 6 constraints, say 3 copies of each or: $A_x = A_y = A_z = B_y = B_z = C_z = 0$, in order to "pin" in the space 3 of the points involved by the couple of constraints, and then we verify that the associated bipartite graph has a perfect matching. We have this time to consider couples of constraints, because we have to pin 3 points and a single constraint may involve only two points. The main idea is that an over-rigid subsystem, having n' points, k' unknowns and more than $P(n', k')$ constraints, contains at least such a couple of constraints; after copying each of them 3 times, we will obtain a partly over-constrained system of equations, which has not a perfect matching. Conversely, if the system fulfills P, then all bipartite graphs will have a perfect matching.

Unfortunately, the P property is not strong enough in $3D$: some graphs fulfill the P property but are not rigid and G. Laman's theorem does not extend to $3D$ and beyond. Fig. 10 shows the classical counterexample seen in [Hen92b, LP86] and others. To not be abused by such configurations, a solution is to also verify (apart the P property) that each time a subset of points is determined by two subsets of constraints A and B, then it is determined by $A \cap B$: here the two "pole vertices" are fixed by the left and by the right halves of constraints, the intersection of which is empty. We are not presumptuous enough to think this condition is always sufficient.

In $3D$ the graph-based approach may seem less attractive, relatively to the numerical probabilistic one which always works, in time $O(n^3)$, for any dimension, and which is straightforward to implement. However, the graph-based approach will give in a natural way the subrigid parts contrarily to the numerical probabilistic approach and deserves further study.

Choosing the Best Solution Amongst Many

After the work of C. Hoffmann, R. Paige and their students [BFH+95], identifying the good solution of a system is now considered as an important issue of this modeling scheme. There are several approaches.

With the numerical methods, the initial guess is supposed to indicate the hoped solution, at least theoretically.

Otherwise, one idea is to choose the solution that keeps better the relative location of geometric elements in the initial guess (for instance the clockwise or anti-clockwise orientation of triplets of construction points). If this heuristic method fails, the decomposition of the set of constraints into basic problems permits [BFH+95] to interactively browse the set of all solutions, considered as a tree of choices: for instance, an equation of degree 2 has two sons, one with the positive square root, the other with the negative square root.

The last possible method is to add inequalities which reject all but the wanted solution. Unfortunately inequalities are difficult to debug.

Conclusion

This chapter has presented two techniques for the qualitative study of systems of constraints: the first, the numerical and probabilistic one, is easy to implement and very general: it works in any dimension. The second stems from graph theory, and generalizes and simplifies previous decomposition approaches. In 2D, it is easy to implement, faster than the first approach, and moreover it gives all rigid subparts of the sketch, which permit to speed up the resolution step. Anyway, it is worth combining the two approaches: the graph-based method gives a first decomposition, each part of which may then be studied further with the numerical and probabilistic technique. Further work is needed for the graph-based approach in $3D$, which has not been implemented and tested so far.

References

[AAJM93] S. Ait-Aoudia, R. Jegou, and D. Michelucci. Reduction of constraint systems. In *Compugraphic*, pages 83–92, Alvor, Portugal, 1993. Also available at *http://www.emse.fr/~micheluc/*.

[AHU83]   A. Aho, J. Hopcroft, and J. Ullman. *Data Structures and Algorithms*. Addison-Wesley Publishing Company, Reading, Mass., 1983.

[AM95]   R. Anderl and R. Mendgen. Parametric design ands its impact on solid modeling applications. In *Proceedings of the Symposium on Solid Modeling Foundations and CAD/CAM Applications*, pages 1–12, 1995.

[BFH⁺95] W. Bouma, I. Fudos, C. Hoffmann, J. Cai, and R. Paige. Geometric constraint solver. *Computer-Aided Design*, 27(6):487–501, 1995.

[Brü85]   B. Brüderlin. Using prolog for constructing geometric objects defined by constraints. In *European Conference on Computer Algebra*, pages 448–459, 1985.

[Brü86]   B. Brüderlin. Constructing three-dimensional geometric objects defined by constraints. In *Interactive 3D Graphics*, pages 111–129, October 1986.

[Brü88]   B. Brüderlin. Automatizing geometric proofs and constructions. In Springer Verlag Lectures Notes in Computer Science, 333, editor, *Computational Geometry and its Applications*, pages 233–252, 1988.

[But79]   M. Buthion. Un programme qui résout formellement des problèmes de constructions géométriques. *RAIRO Informatique*, 3(4):353–387, oct 1979.

[DMS97]   J.F. Dufourd, P. Mathis, and P. Schreck. Formal resolution of geometrical constraint systems by assembling. In Christoph Hoffmann and Wim Bronsvort, editors, *Fourth Symposium on Solid Modeling and Applications*, pages 271–284. ACM press, 1997.

[Hen92a] B. Hendrickson. Conditions for unique realizations. *SIAM J. Computing*, 21(1):65–84, feb 1992.

[Hen92b] B. Hendrickson. Conditions for unique realizations. *SIAM J. Computing*, 21(1):65–84, feb 1992.

[HK73]   J.E. Hopcroft and R.M. Karp. An $n^{5/2}$ algorithm for maximum matching in bipartite graphs. *SIAM J. Computing*, 2(4):225–231, 1973.

[HM93]   Joos Heintz and Jacques Morgenstern. On the intrinsic complexity of elimination theory. Technical Report 1923, INRIA, 1993.

[Kra91]   G.A. Kramer. Using degrees of freedom analysis to solve geometric constraint systems. In *Proceedings of the Symposium on Solid Modeling Foundations and CAD/CAM Applications*, pages 371–378, 1991.

[Lam70]   G. Laman. On graphs rigidity of plane skeletal structures. *J. Engineering Math.*, 4(4):331–340, Oct. 1970.

[LM96]   H. Lamure and D. Michelucci. Solving geometric constraints by homotopy. *IEEE Trans. Visualization and Comp. Graphics*, 2(1):28–34, 1996.

[LP86]   L. Lovasz and M.D. Plummer. *Matching Theory*. North-Holland, 1986.

[Mar71]   W.A. Martin. Determining the equivalence of algebraic expressions by hash coding. *J. ACM*, 18(4):549–558, 1971.

[Owe91]   J.C. Owen. Algebraic solution for geometry from dimensional constraints. In *Proceedings of the Symposium on Solid Modeling Foundations and CAD/CAM Applications*, pages 397–407, 1991.

[Rec86]   A. Recski. *Matroid theory and its applications*. Springer Verlag, 1986.

[Sch80]   J.T. Schwartz. Fast probabilistic algorithms for verification of polynomial identities. *J. ACM*, (27):701–717, 1980.

[Tar72]   R.E. Tarjan. Depth first search and linear graph algorithms. *SIAM J. Computing*, 1, 2:146–160, 1972.

[VSR92]   A. Verroust, F. Schonek, and D. Roller. Rule oriented method for parametrized computer aided design. *Computer Aided Design*, 24(3):531–540, Oct 1992.

Geometric Relaxation for Solving Constraint-Based Models

Lluis Solano Albajes, Pere Brunet Crosa [1]

In this chapter, iterative algorithms for solving geometric constraint-based models are discussed, and a new relaxation algorithm is proposed. The algorithm starts from a constructive symbolic representation of objects (Constructive Parametric Solid Model, CPSM) and it proceeds by iterative relaxation of the geometric constraints. Systems that can be reduced to distance and angle constraints can be handled. The convergence of the proposed algorithm is discussed and an example is presented and discussed.

Introduction

The term CAD is related to the use of computers as an aid to the entire design process, including creation, modification, and visualization of designed parts. Traditional CAD-systems focus on the explicit object being designed. The construction process and the relations between the involved elements are not reflected in the final design.

Constraint-based modeling allows the user to define object families (generic objects) that can be subsequently converted onto specific objects by giving the values of the generic object parameters and solving the defined constraints.

Design process using CAD-systems is interactive: the user specifies step-by-step a sequence of operations that converges towards the final object. User interaction is performed through a Graphical User Interface. Modeling operations and geometric constraints can be choosen by the user

---

[1]Departament de Llenguatges i Sistemes Informàtics. Universitat Politècnica de Catalunya. Spain

in order to define the object. With the use of geometric constraints it is possible to specify the organization of a design. In this sense it is easy to generate design variations using constraints [Rol91]. Constraint-based design is aimed at representing and capturing designer's intent. It alows one to design generic objects rather than explicit ones and ever a family of designs instead a single one.

The proposed constraint representation is based on the Constructive Parametric Solid Model (CPSM) that is a procedural description of the modeling operations sequence and geometric constraints performed by the user during the interactive object design [SoB93].

The CPSM is the representation of a generic object of the whole family of objects, and keeps the incremental design process. Previous defined models can be instantiated when a design is in progress; 2D, 3D and 2D_to_3D operations are supported. The CPSM is an Editable Representation (EREP) suitable for storing and transmission, it supports both generic and specific designs, and records the conceptual construction steps [RBN89] [HoJ92]. Each statement represent a modeling operation or geometric constraint, expressed in terms of a definition language [SoB94a]. In order to manage and keep the geometric constraints a specific structure called the Internal Model Representation is used.

The Internal Model Representation

The Internal Model Representation (IMR) is the structure that we use to manage and represent geometric constraints [SoB94b]. In the design process, the user can define dimensional constraints (distances and angles) between existing geometric elements. The IMR keeps explicit constraints, implicit constraints and default constraints that can be defined while the design is in progress. Constraint satisfaction uses the information stored in the IMR.

More precisely, we can define,

Internal Model Representation as a graph I that keeps and manages constraints R between the existing geometric elements G. Nodes N in the graph are points and vectors involved in the defined constraints and edges C are constraints that relate nodes.

Thus we have:
$$I = graph\langle N, C\rangle$$
where N is the set of nodes that represent points or vectors on G and C is a set of edges that represent constraints between nodes. $C \subseteq N \times N$

If p is a point of the model:
$$\forall p \in G \text{ if } \exists c \in R \mid c(p) \Rightarrow p \in N$$

A graph that represents the existing constraints between a set of elements is called *constraint graph*. All the modeling operations and constraints that are possible in the CPSM [SoB93] can be represented as distances between two points and angles between two vectors, see [SoB98]. Geometric constraints in CPSM can be therefore translated into distances and angles.

In a CPSM model with n points 2D, we say that is well constrained iff there are $2n - 3$ constraints and there doesn't exist any subgraph G' with n' with more than $2n' - 3$ [Lam70].

It has also been proved [SoB98] that if a model is well-constrained it is always possible to express the angle constraints as a function of a set of distance constraints. As a consequence the IMR constraints are translated only to distance constraints. The IMR keeps the explicit constraints that have been defined by the user, the implicit constraints that reflect relationships between geometric elements, for example a point that belongs to a polygon, and default constraints that reflect the dimensions of the starting design of the product. The solver deals with this kind of constraints in a iterative way.

Geometric Constraint Solver

In the literature there are different approaches to solve geometric constraints. There are two main approaches: the constructive approach and the equational approach. In the constructive approach, constraints are satisfied constructively. The geometric elements are placed in some order. Constructive methods are classified in:

- Procedural methods [Rol91a] [Rol91b] [RBN89] [CFV88] [Emm90] [SoB94b]

- Graph based methods [Owe91] [BFHCP93] [FuH93]

- Rule based methods [Ald88] [Sun88] [SoB91c] [VSR92] [JoS97]

In equational methods, constraints are translated to equations. The system of non-linear simultaneous equations is solved using different techniques such as:

- Newton-Raphson [LGL81] [Nel85] [SoB93]

- Geometric iteration [Bor81] [HsB97]

- Symbolic algebraic methods, as Gröbner basis [Buch88] or Wu-Ritt [Wen86] method

- Propagation methods [Sut63] [StS80]

In order to solve the distance constraints also it is possible to work in terms of energy minimization [WFB87] [KaB90]. In this case, constraints are relaxed in order to decrease the energy level using gradient techniques. The main problem is that the system can fall in local minima.

Iterative approach proposed

Before introducing the approach that we are working it is necessary to introduce some basic definitions. In this definitions we assume the planar case with n points $P_1, \ldots, P_n$

Solver definitions

State Vector $\vec{d}$: it contains actual values of the parameters involved with the existing m constraints. $dim(\vec{d}) = m$

$$\vec{d} = \left[\begin{array}{c} d_1 \\ \vdots \\ d_m \end{array} \right]$$

where d_k is a distance constraint between its endpoints P_{k_1}, P_{k_2}; $k_1, k_2 \in K, \sharp K = n, m = 2n - 3$ in the plane.

Equilibrium Vector $\vec{d}^c$: it constains the target values of the parameters involved with the m constraints defined. $dim(\vec{d}^c) = m$

$$\vec{d}^c = \begin{bmatrix} d_1^c \\ \vdots \\ d_m^c \end{bmatrix}$$

Difference Vector $\vec{\delta}$: it is the difference between state vector $\vec{d}$ and equilibrium vector $\vec{d}^c$.

$$\vec{\delta} = \vec{d} - \vec{d}^c$$

Constraint Variation Matrix $A = [a_{ji}]$: each matrix element represents the differential variation of the parameter associated with constraint j according to the variation of the constraint i $(i \neq j)$. So, a_{ji} represents, the differential variation of the parameter d_j according to the variation of d_i.

$$dim(A) = m \cdot m$$

$$A = \begin{cases} a_{ji} = 1 & \text{if } i = j \\ a_{ji} = 0 & \text{if } i \neq j \text{ and } d_i, d_j \text{ doesn't share a common endpoint} \\ a_{ji} = \frac{\partial d_j}{\partial d_i} & \text{if } i \neq j \text{ and } d_i, d_j \text{ share a common endpoint} \end{cases}$$

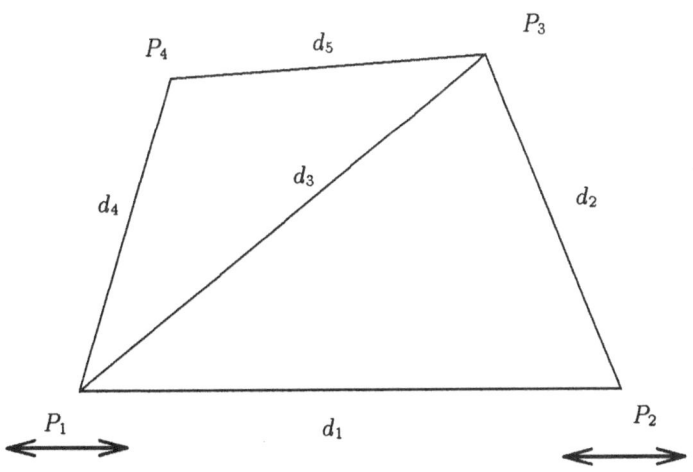

Figure 1: Link between distance constraints

The matrix column k represents the unit variation of the overall constraints that are linked with d_k. For instance, in figure 1 the variation of d_1 changes the location of P_1 and P_2 and also the value of d_2, d_3, d_4. As a consequence a_{21}, a_{31}, a_{41} are different from 0, but the value of d_5 doesn't change and $a_{51} = 0$. Moreover as the solver works with distance constraints the constraint variation matrix is symmetric and it is easy to prove that $|a_{ji}| \leq 1$.

It is possible to describe the evolution of the proposed method in terms of the elements defined previously, provided that r is small:

$$\vec{\delta}]_{\kappa+1} = \vec{\delta}]_\kappa + A_\kappa \cdot R_i \cdot (\vec{\delta}^c - \vec{\delta}]_\kappa)$$

where $\vec{\delta}]_\kappa$ is the value of the difference vector $\vec{\delta}$ at the iteration κ and R_i is a matrix that shows which constraint d_i is solved $(dim(R) = m \cdot m)$.

$$R_i = \left\{ \begin{array}{ll} R_{ii} = r & \\ R_{jk} = 0 & \text{if } j \neq i \text{ or } k \neq i \end{array} \right.$$

Constraint Solver

Geometric relaxation solvers can work by iterating one of the following cases:

- One point relaxation: On a point P, if the $grad(\vec{\delta}) \neq 0$, we can move P in the direction of the gradient vector. In this way we have maximum disminution of $||\vec{\delta}||$. A problem raises when there is a local minimum and the gradient vector is null for all points.

- One constraint relaxation: It is also possible to relax one constraint d_i. In this case we can also fall into local minima. If the constraint relaxed d_i has the maximum value of $|\vec{\delta}|$ we have experimentally shown a good behaviour of the convergence of the solver.

- Global point relaxation: In this case a space deformation can be applied in order to preserve the decreasing of $||\vec{\delta}||$. It can be shown [SoB98] that it requires solving a linear system of dimension m at each iteration. On the other hand convergence is guaranteed [SoB98].

- Global constraint relaxation: This case would required to solve all the constraints at the same time. Therefore, it is unaffordable.

In this context, we are working on a constraint solver based on an iterative relaxation method. While $||\vec{\delta}|| \geq \varepsilon$, at each iteration, the constraint solver works on points or constraints. This means that the solver in each step has to decide to select one constraint to relax or to perform a space deformation in order to move slightly all the points.

In the iterative constraint solver that we are developing, the relaxation process is performed in the following way.

Constraint relaxation. At each iteration one constraint or the whole set of points is deformed. Our algorithm works at each iteration in the following way:

If the system is not in a local minimum status then

- A constraint i is choosen such that $|\delta_i| \geq |\delta_j| \; \forall j$

- relaxe constraint i ($d_i \rfloor_\kappa$) by making it closer to the equilibrium value d_i^c. A relaxation factor r is used.

$$d_i \rfloor_{\kappa+1} = d_i \rfloor_\kappa + r \cdot (d_i^c - d_i \rfloor_\kappa)$$

- update constraints d_j such that $a_{ji} \neq 0$

$$d_j \rfloor_{\kappa+1} = d_j \rfloor_\kappa + r \cdot a_{ji} \cdot (d_i^c - d_i \rfloor_{\kappa+1})$$

- update geometric location of points affected by the constraint modification values.

- compute the new matrix values of A and $\vec{\delta}$.

Else

- A global point deformation is used in order to move the whole set of points and decrease $||\vec{\delta}||$.

The global deformation (for more details see [SoB98]) works by converting the constrained system to a representation based on the set of relative vectors defined by

$$\vec{R}_i = \frac{P_{i2} - P_{i1}}{d_i^c}$$

P_{i1} and P_{i2} being the endpoints of the distance constraint d_i. Using this representation, it can be shown that it is always possible to compute a displacement of all points P_j of the system in a way such that

$$||\vec{R_i}|| \to 1 \; \forall i$$

The displacement computation requires solving a linear system of equations at each iteration. In the limit, the relaxation algorithm produces $||\vec{R_i}|| = 1 \; \forall i$ and therefore the solver converges.

In figure 2, an example involving six points and nine constraints is shown. The distance constraints are:

$$(P_1 P_2) \; (P_2 P_3) \; (P_3 P_4) \; (P_4 P_5) \; (P_5 P_6) \; (P_6 P_1) \; (P_1 P_4) \; (P_2 P_5) \; (P_3 P_6)$$

The figure shows the initial configuration and the evolution of the relaxation, showing two intermediate states. We have experimentally confirmed the local minima are successfully avoided using the proposed algorithm.

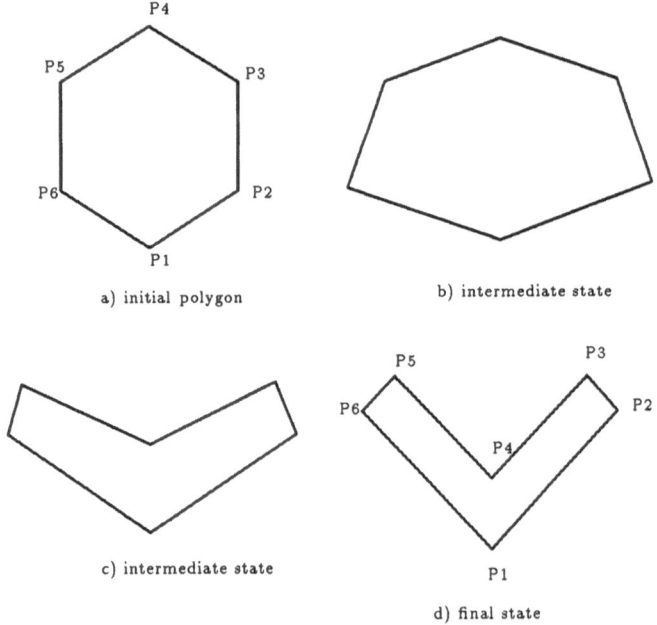

Figure 2: Example of the solver evolution

Conclusions

We have presented a constraint solver that works on the Internal Model Representation and which is based only on distance constraints. Through a iterative relaxation of constraints a solution is found. The convergence of the solver using constraint and point relaxation can be guaranteed.

The proposed method is applied in 2D but the method is extensible to 3D. The relaxation process starts from the initial conditions. In this way we provide a mechanism to reach a solution which is, in some sense, close to the initial conditions. In cases with several solutions, by changing the initial conditions or the relaxation parameter r, it is possible to switch from one solution to another. The solver can also be applied in under-constraint systems keeping the user intent.

Future work will involve the analysis of suitable values for the relaxation factor and constraint consistency

References

[Ald88]   B.Aldfeld. Variation of geometries based on a geometric-reasoning method. CAD, vol.20, no.3, April 1988.

[Bor81]   A.H.Borning. The programming language aspects of ThingLab, a constrained oriented simulation laboratory. ACM Trans. on Prog. Lang. and Systems, vol.3, no.4, October 1981.

[BFHCP93]   W.Bouma, I.Fudos, C Hoffman, J.Cai, R.Paige. A geometric constraint solver. Technical Report, CSD-TR-93-054, Purdue University, 1993.

[Buch85]   B.Buchberger. Gröbner Bases: An algorithmic Method in Polynomial Ideal Theory. In N.K.Bose, editor, Multidimensional Systems Theory, pp.184-232. D.Reidel PublishingCo., 1985.

[CFV88]   U.Cugini, F.Folini, I.Vicini. A Procedural System for the Definition and Storage of Technical Drawings in Parametric

Form. Proceedings of Eurographics'88, pp.183-196, North-Holland, 1988.

[Emm90]   M.J.G.M.van Emmerick. Interactive Design of Parameterized 3D models by direct manipulation. PhD thesis, Delft University Press, 1990.

[FuH93]   I.Fudos, C.Hoffman. Correctness proof of a geometric constraint solver. Technical Report CSD-93-076, Computer Sciences Departament, Purdue University, December 1993.

[HoJ92]   C.M.Hoffmann, R.Juan. ERep. An editable high-level representation for geometric design and analysis. Technical Report CSD-TR-92-055. CAPO Report CER-92-24. Purdue Unversity, August 1992.

[HsB97]   C.Hsu, B.D.Brüderlin. A Hybrid Constraint Solver Using Exact and Iterative Geometric Constructions. In CAD Systems Devolopment: Tools and Methods, D.Roller and P.Brunet Eds, Springer Verlag 1997, pp. 265-279

[JoS97]   R. Joan-Arinyo, A. Soto. Rule-Based Geometric Constraint Solver. Computer & Graphics, vol.21 no.5, 1997.

[KaB90]   D. Kalra, A.H. Barr. A Constraint-Based Figure-Maker. Proceedings of Eurographics'90, pp.413-424, North-Holland, 1990.

[Lam70]   G.Laman. On Graphs and Rigidity of Plane Skeletal Structures. Journal of Engineering Mathematics, vol.4, no.4, October 1970.

[LGL81]   V.C.Lin, D.C.Gossard, R.A.Light. Variational Geometry in Computer Aided Design. ACM Computer Graphics, vol.15, no.3, August 1981.

[Nel85]   G.Nelson. Juno, a constraint-based graphics system. SIGGRAPH'85, vol.19, no.3, pp.235-243, San Francisco. July 22-26, 1985.

[Owe91]   J.C.Owen. Algebraic Solution for Geometry from Dimensional Constraints. Proceedings of Symposium on Solid Modelling Foundations and CAD/CAM Applications. J.Rossignac, J.Turner (eds). Austin, June 5-7, pp.397-407, ACM Press 1991.

[RBN89]   J.R.Rossignac, P.Borrel, L.R.Nackman. Interactive Design with Sequences of Parameterized Transformations. Intelligent CAD Systems II. V.Akman, P.J.W.ten Hagen, P.J. Veerkamp (eds), pp.93-125, Springer-Verlag, 1989.

[Rol91a]   D.Roller. Advanced Methods for Parametric Design. Geometric Modelling. Methods and Applications. H.Hagen, D.Roller (eds), pp. 251-266, Springer-Verlag, 1991.

[Rol91b]   D.Roller. An approach to computer-aided parametric design. CAD, vol.23, no.5, June 1991.

[SoB91c]   W. Sohrt, B.D.Brüderlin. Interaction with constraints in 3D modeling. International Journal of Computational Geometry & Applications, vol.1, no.4, pp.405-425, 1991.

[SoB93]   L.Solano, P.Brunet. A system for constructive constraint-based modelling. In B.Falcidieno and T.Kunii, editors, Modeling in Computer Graphics. Springer Verlag.

[SoB94a]   L.Solano, P.Brunet. A language for constructive parametric solid modelling. Technical Report LSI-94-43-R, Universitat Politecnica de Catalunya, LiSI, 1994.

[SoB94b]   L.Solano, P.Brunet. Constructive constraint-based model for parametric CAD systems. CAD, vol.26, no.8, pp.614-621, August 1994.

[SoB98]   L.Solano, P.Brunet. A geometric constraint solver for Parametric Constraint-Based Models. Technical Report LSI-98, Universitat Politecnica de Catalunya, LiSI, 1998.

[StS80]   G.L.Steele, G.L.Sussman. Constraint - a language for expressing almost-hierarchical descriptions. Artificial Intelligence, pp.1-39, January 1980.

[Sun88]   G.Sunde. Specification of Shape by Dimensions and Other Geometric Constraints. Geometric Modelling for CAD Applications. M.J.Wozny, H.W.McLaughlin, J.L.Ecarnacao (eds), pp.199-213, North-holland 1988.

[Sut63]   I.Sutherland. Sketchpad, a man-machine graphical communication system. In Proc. of the Spring Joint Comp. Conference, pp.329-345. IFIPS, 1963.

[VSR92]   A.Verroust, F.Schonek, D.Roller. Rule-oriented method for
           parametrized computer aided design. CAD, vol.24, no.10,
           October 1992.

[Wen86]   Wu Wen-Tsün. Basic principles of mechanical theorem
           proving in geometries. Journal of Systems Sciences and
           Mathematical Sciences, vol.4, pp.207-235, 1986.

[WFB87]   A. Witkin, K.Fleischer, A. Barr. Energy Constraints on Pa-
           rameterized models. Computer Graphics, vol.21, pp.225-
           232, 1987.

Chapter 4

Constraints for Freeform Surfaces

Overview

Variational Design and Surface Reconstruction

Hans Hagen, [1] Siegfried Heinz, [2] Michael Thesing, [3] Thomas Schreiber [4]

Curves and surfaces designed in a computer graphics environment have many applications, including the design of cars, airplanes, shipbodies and modelling robots. These free-form objects are an essential part of powerful CAD-systems.

A major topic is "reverse engineering", an important part in this content is the surface reconstruction problem. This means a set of points delivered from a scanning system must be processed to a surface model, that can be handled in a CAD/CAM system.

This paper describes the structuring of point data as a pre-processing step before the actual surface reconstruction takes place. First the focus is set to an algorithm to reduce the data of large point sets. Subsequenty, a method for curvature approximation, based on the Delaunay triangulation of a set of points is described. By means of curvature discontinuities the points are grouped to represent separate surface-patches of the model. As a final step a special variational design technique can be used to generate the surface.

Introduction

The process to create a 3D CAD model from an existing physical model is called reverse engineering, as opposed to normal engineering, where a physical model is created from a CAD model. This paper proposes a five step solution for that process. The first step is digitizing or measuring

---

[1] University of Kaiserslautern, Germany
[2] CAE-Beratung, Germany
[3] TransCAT GmbH, Germany
[4] University of Kaiserslautern, Germany

the physical model, merging the data of different views into one large set of scattered points, reducing these points results in a cloud of points which describe the object sufficiently. Proceeding from this data the next step is to establish a local neighbourhood of each point which can be done by a triangulation of the points. The third step is the segmentation of the points. Each group of points has to represent a separate surface of the final model. The groups are determined by means of curvature discontinuities which are approximated for each point based on the neighboured points. Afterwards the fourth step is the final surface reconstruction of each group of points. The free form surfaces are generated with variational design methods considering earlier created surfaces to get point and tangent continuity to adjacent surfaces. The other analytical surfaces like planes, cylinders and spheres are created using standard CAD tools. To conclude the final CAD model the generated surfaces have to be analyzed to validate the quality of the surfaces for further requirements. This last step has to guarantee the quality of each surface (e. g. curvature distribution) and the quality of the whole CAD model (e. g. tangency between two adjacent surfaces). That implies an iteration for the fourth and fifth step of the reverse engineering process. After the iterations the CAD model is ready to be used for finite element analysis or NC programming.

To reduce a large set of unstructured point data with a minimum lack of information a method proposed by Schreiber (see [Sch94]) is used in chapter 1. In the next chapter a Delaunay triangulation of the points to determine if two points are adjacent to each other is explained. This triangulation is based on a two dimensional case of the general algorithm that Schreiber describes in his dissertation [Sch94]. For the curvature estimation at a point p, Hamanns approach (see [Ham91]) was extended by using a general polynomial function to approximate a local set of points around the point p. For a segmentation of the points using the estimated curvatures two methods will be introduced in chapter 3. One method uses a threshold value for the minimal curvature and the second method searches for normal and curvature continuities. Chapter 4 extends the variational design principle for surface reconstruction, proposed by Hagen/Santarelli (see [Hag92],[San94]), which combines a weighted least square approximation with an automatic smoothing process. This principle was extended to use any degree and any continuity between segments of the surface and to consider adjacent surfaces for point and tangent continuity to the adjacent surfaces. Furthermore, the point parametrization will be included in the variational design process.

Digitizing - Data Improvement - Data Reduction

Physical objects can be digitized using manual as well as CNC-controlled Coordinate Measuring Machines (CMM) or laser range scanning systems. Using laser range scanning systems, most of the objects have to be scanned from different views and the different views have to be merged into one large set of points. To be independent from a chosen measuring method or measuring machine the algorithms below are using only large sets of arbitrary distributed points, no structuring of these points is needed. To minimize the amount of point data, a cloud of points is calculated using data improvement and data reduction methods, that sufficiently describes the object.

Let $P = \{p_i \in \mathbb{R}^3 | i = 1, \ldots, n\}$ be a set of n distinct points. To reduce P to a smaller set $Q = \{q_j \in P | j = 1, \ldots, m < n\}$, a subdivision into m distinct clusters can be calculated.

That clustering means grouping similar points by optimizing a certain criterion function in a way that a subdivision results as naturally as possible. Subsequently, a single point out of each cluster has to be selected, that represents the group of points belonging to the cluster. These chosen points build the reduced point set Q. As criterion to verify the quality of the subdivision a function k_h can be introduced, that assigns a numerical value to each cluster C_h.

$$k_h = \sum_{\substack{i,j \in I_h \\ i < j}} \|p_i - p_j\|^2$$

with $I_h = \{i | p_i \in C_h\}$ and $h = 1, \ldots, m$.

This cost of a cluster is a measure of the distribution of the points in the cluster. An optimal subdivision of P is given by minimizing the costs. That is

$$\sum_{h=1}^{m} k_h = \sum_{h=1}^{m} \sum_{\substack{i,j \in I_h \\ i < j}} \|p_i - p_j\|^2 \rightarrow min$$

This expression is equivalent to

$$\sum_{h=1}^{m} \sum_{i \in I_h} \|p_i - S_h\|^2 \to min; \qquad\qquad S_h = \frac{1}{|C_h|} \sum_{i \in I_h} p_i$$

where the point S_h is called the center of the cluster C_h.

The problem to find a global minimum of this expression, and so a global segmentation of the points, is known to be np-complete. Therefore a heuristic method is used to find a suboptimal solution. The described method is based on an iterative refinement of an initial subdivision.

The initial subdivision is the single cluster containing all points of P. In each step the cluster with the highest costs will be determined and divided into two new clusters, so that the costs are locally optimal reduced. For this purpose a proper hyperplane has to be calculated which may be defined perpendicular to the direction of the largest eigenvector of the covariance matrix of the points in a cluster.

If the set P is divided into m clusters, one knot point p_h from the points of each cluster has to be selected. The optimal one would be S_h, so the point nearest to the centre is used as the knot point. The knot point of the m clusters defines the smaller point set Q. (For further details see [Sch91].)

Triangulation

The second step in the reverse engineering process has to determine the local neighbourhood of each point and can be done by a triangulation of the points. In the following an iterative algorithm will be presented to construct a two dimensional Delaunay triangulation of a point set P. The construction algorithm inserts one point after the other into an initial triangulation, which is extended appropriately in each step.

First of all the points of P have to be projected into one plane. The best approach is to calculate the mean plane (best fitting plane in the least square sense) of the points. Another possibility is to let the user choose a proper plane. If no plane on which the points can be projected on properly 3D surface triangulation methods have to be used (see [Boi84], [DeR92], [Ede94], [Sch97]). In the following we assume the existance of

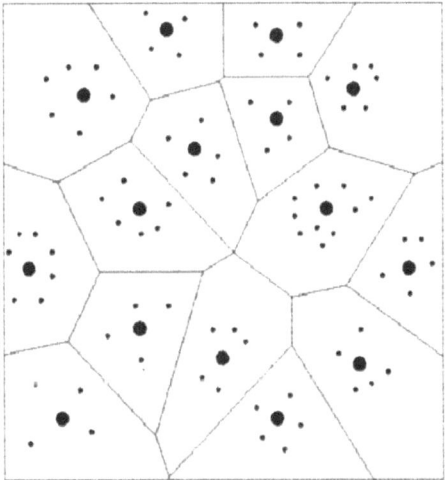

Figure 1: Voronoi diagram of a point set (small dots)with its cluster center points (bold dots).

a proper projection plane.

An obvious choice for the initial triangulation is a single triangle, that consists of the first three points of P. Nevertheless a different method will be presented. To determine the initial triangulation a circumscribed rectangle of the points in P can be calculated. The initial triangulation consists of the two triangles, that are derived from this rectangle and any of its diagonals. The advantage of this initial triangulation is, that every point of the set P is interior.

To insert a point $p \in P$ during one step of the iteration, the procedure is as follows. First search the triangle, that contains the point p. This can be done by an oriented walk. It starts with any triangle and checks if p is interior. To check if a point is interior to a triangle the position of p according to each of the three edges has to be determined. With the assumption that all triangles are positively oriented the following criterion can be used.

Then calculate the determinants for the three edges (p_1, p_2), (p_0, p_2) and (p_0, p_1) of a triangle defined by the three vertices p_0, p_1, p_2:

$$d_1 := det \begin{vmatrix} p & p_1 & p_2 \\ 1 & 1 & 1 \end{vmatrix}, d_2 := det \begin{vmatrix} p_0 & p & p_2 \\ 1 & 1 & 1 \end{vmatrix}, d_3 := det \begin{vmatrix} p_0 & p_1 & p \\ 1 & 1 & 1 \end{vmatrix}$$

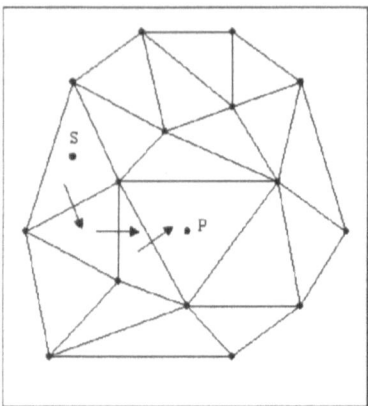

Figure 2: Oriented walk

If a determinant of an edge is negative then p is called invisible, otherwise it is called visible. If p is visible from all three edges, i. e. all determinants are positive, p is interior to the triangle and the search stops. Otherwise p is exterior and the algorithm continues with the triangle next to the edge with a negative determinant. To extend a Delaunay triangulation by a point p, the following must be considered. A Delaunay triangulation has the so-called circumcircle property: no point of a Delaunay triangulation is interior to the circumcircle passing through the three vertices of a triangle. Therefore all triangles that violate this property after insertion of p must be deleted. They are replaced by triangles that embed the point p into the triangulation. This can be done using two lists: the list DLL, containing all triangles that violate the criterion and the list NDL, containing the triangles replacing them.

The triangles of the list DLL are exactly those triangles with a circumcircle containing p. The triangles containing the point p, determined by the oriented walk algorithm, belong obviously to the list DLL. Starting with one triangle, the circumcircles of all adjacent triangles have to be calculated. If p is interior to a circle, the procedure adds the belonging triangle to the list DLL and continues recursively with this triangle. The procedure stops if p is exterior to all circumcircles.

To create the list NDL, all edges of each triangle in the list DLL have to be checked for the existence of adjacent triangles. If there are adjacent triangles not within DLL list, new triangles have to be calculated based on the point p and the common edges. These triangles are added to the list NDL.

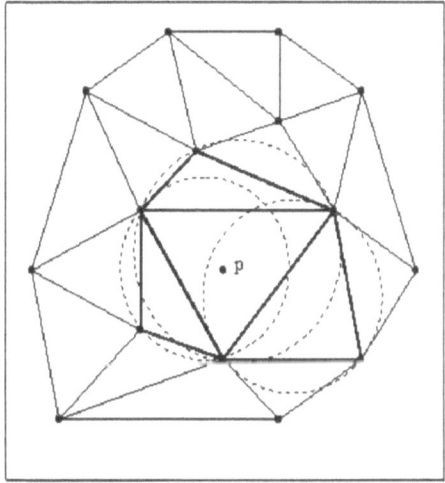

Figure 3: Triangles of list DLL

The triangulation is finished when all triangles of the list DLL are removed and all triangles of NDL are inserted.

Curvature and discontinuity based segmentation

For a realistic segmentation of the points, all points have to be grouped in a way, that each group of points represents one surface of the final CAD model. With the following new method of curvature approximation, based on the Delaunay triangulation explained above, the points can be grouped using two methods. For both methods it is essential to approximate the curvature in each point of the triangulation.

The curvature at a point p can be estimated by calculating an approximating function f for a local set of points around p. Then the curvature at p can be set to the curvature of the approximating function in p.

In the dissertation of Hamann [Ham91] the so-called osculating paraboloid is used as the approximating function. It has the canonical form (surface is tangent to the xy plane at the origin)

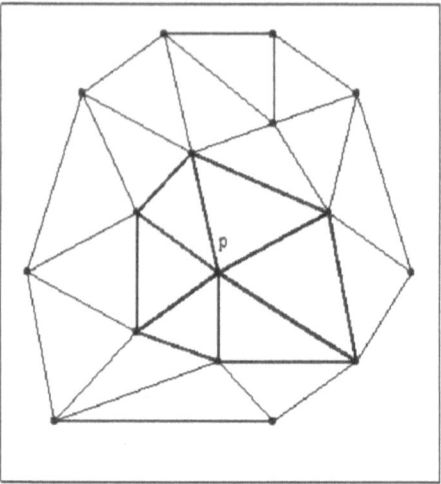

Figure 4: Triangles of list NDL

$$f(x,y) = \frac{1}{2}(c_{2,0}x^2 + 2c_{1,1}xy + c_{0,2}y^2)$$

and interpolates the point p while approximating the neighbouring points. This approach was extended by using a general polynomial function f to approximate a local set of points around p:

$$f(u,v) = \sum_{s=0}^{deg_u} \sum_{t=0}^{deg_v} \vec{c}_{s,t} u^s v^t$$

First a set of points neighbouring p have to be determined. For further reference this set of points is called the platelet and the points are called platelet points. The platelet consists of all points that share a common edge of a triangle with p. The platelet can be extended adding all points that share a common edge with any platelet point. For a better curvature estimation this extension can be repeated several times. In spite of this, for the detection of discontinuities in the point cloud, it is essential to keep the approximation as local as possible.

After determining the platelet points the approximating function f has to be parametrized. The simplest method is to use a plane defined by the two directional vectors $\vec{v}_1, \vec{v}_2$, and the origin $\vec{u}_e$, that resign to the following parametric representation:

$$\vec{x} = \vec{v}_1 \cdot u + \vec{v}_2 \cdot v + \vec{u}_e$$

Then the parameter values (u_j, v_j) of the platelet points $\vec{y}$ according to this plane can be computed:

$$u_j = \vec{v}_1 \cdot (\vec{y}_j - \vec{u}_e), \quad v_j = \vec{v}_2 \cdot (\vec{y}_j + \vec{u}_e); \quad j = 0, \ldots, n_i$$

The constraints of approximating the platelet points by the function f result from the distances of the points to this function. A least squares approach minimizes the sum of the squared distances:

$$SL := \sum_{j=0}^{n_i} (D_j)^2 = \sum_{j=0}^{n_i} (f(u_j, v_j) - \vec{y}_j)^2 \to min.$$

The necessary conditions for a minimum of this sum is given by

$$\frac{\partial}{\partial \vec{c}_{q,r}} SL = 0 \implies \sum_{j=0}^{n_i} f(u_j, v_j) \cdot u_j^q v_j^r = \sum_{j=0}^{n_i} \vec{y}_j \cdot u_j^q v_j^r; \quad \begin{array}{l} q = 0, \cdots, deg_u \\ r = 0, \cdots, deg_v \end{array}$$

These leads to a linear system of equations with $(deg_u + 1) \cdot (deg_v + 1)$ vectorial unknowns. This equation system is only solvable, if the platelet contains at least as many points as unknowns. Proper methods to solve this system can be found in the standard literature of numerical computation. The final computation of the curvature in the point $f(u_0, v_0)$ can be done by solving the following quadratic equation:

$$k^2 - 2\frac{LG - 2FM + EN}{2(EG - F^2)}k + \frac{LN - M^2}{EG - F^2} = 0$$

E, F, G, L, M, N are the first and the second fundamental coefficients of a surface with the parametric representation $\vec{X}(u, v)$ and the unit normal vectors $\vec{N}(u, v)$:

$$E = \vec{X}_u \cdot \vec{X}_u, \quad F = \vec{X}_u \cdot \vec{X}_v, \quad G = \vec{X}_v \cdot \vec{X}_v$$
$$L = \vec{X}_{uu} \cdot \vec{N}, \quad M = \vec{X}_{uv} \cdot \vec{N}, \quad N = \vec{X}_{vv} \cdot \vec{N}$$

The two roots of this equation are the principal curvatures. In the following only the minimal curvature is used.

Two methods to group the points of a set P using these estimated curvatures can be presented. The first method uses a threshold value λ for the minimal curvature. The points will be separated in different groups, depending if their minimal curvature is greater or less than the threshold value. So the set P is subdivided into groups that have the curvature intervals $]-\alpha, \lambda]$ and $]\lambda, +\alpha[$. This subdivision can refined by selecting a series of further threshold values and applying the method to the already created groups. Therefore groups arise with smaller and smaller curvature intervals.

The second method searches for tangent and curvature discontinuities of neighbouring points. It only attaches two adjacent points to the same group, if the deviation of the two normal vectors is smaller than a given tolerance and the difference of the two minimal curvatures does not exceed a second given tolerance. Otherwise it creates different groups for each of the two points.

Variational Design considering boundary conditions

The fourth step in the reverse engineering process is the final surface reconstruction for each group of points. Analytical surfaces like planes, cylinders and spheres can be created using standard CAD tools. Also the fillets with one constant radius in one parameter direction, as a connecting surface between two given surfaces, can be generated with standard CAD tools. So the focus in this chapter is set to free form surfaces. They can be generated with variational design methods and have to consider earlier created surfaces to keep point and tangent continuity to adjacent surfaces. That implicates the possibility for a user to pre-define boundary curves for point continuity and to select neighbouring surfaces for tangent continual transitions.

The variational design process of Hagen and Santarelli (see [Hag 92],[San 94]) combines a weighted least squares approximation with an automatic smoothing of the surface. The chosen smoothing criterion minimizes the variation of the curvature along the parameter lines of the designed surface. This fundamental B-spline approach was extended for arbitrary degree and arbitrary continuity in both parameter directions and considering given boundary informations.

For the design of complex surfaces a lot of user interaction is needed. The CAD user is responsible for the form and the quality of the generated surfaces. Therefore the variational design of surfaces yields two possible user-defined input parameters: order of the surface in u and v direction

- continuity between segments of the surface in u and v direction

- segmentation in u and v direction (i. e. determine the u and v parameter of the segment boundaries)

- smoothing weights

- surface boundary curves, if available from adjacent earlier created surfaces

- adjacent surfaces for tangent continual transition along a boundary, if available and if the boundary is given

For that the following mathematical model can be used as variation principle:

$$(1 - w_s) \left\{ \sum_{k=1} w_{p_k} \left(F(u_k, v_k) - p_k \right)^2 \right\}$$

$$+w_s \left\{ \sum_{i=1}^{n} \sum_{j=1}^{m} \left(w_{u_g} \int_{v_j}^{v_{j+1}} \int_{u_i}^{u_{i+1}} w_{u_{ij}} \left\| \frac{\partial^3 F(u, v)}{\partial u^3} \right\| dudv \right. \right.$$

$$\left. \left. +w_{v_g} \int_{v_j}^{v_{j+1}} \int_{u_i}^{u_{i+1}} w_{v_{ij}} \left\| \frac{\partial^3 F(u, v)}{\partial v^3} \right\| dudv \right) \right\} \rightarrow min.$$

where $F(u, v)$ is the representation of the surface, $\{p_k | k = 1, \ldots, n_p\}$ is the group of points, n and m are the number of segments in u and v direction, and $w_s, w_{u_g}, w_{v_g}, w_{u_{ij}}$ and $w_{v_{ij}} \in [0, 1]$ are the smoothing weights.

This is a convex combination of least square fitting and approximate jerk minimization along parameter lines. This principle can be applied to B-spline surfaces of the order $p, q \geq 1$ and the continuity of the segments $c_u, c_v \geq 0$

$$F(u, v) = \sum_{i=1}^{ncu} \sum_{j=1}^{ncv} d_{ij} N_i^p(u) V_j^q(v)$$

with ncu, ncv number of control points and the knot vectors

$$U := \left\{ \underbrace{u_1, \ldots, u_1}_{p}, \underbrace{u_2, \ldots, u_2}_{1_u}, \ldots, \underbrace{u_n, \ldots, u_n}_{1_u}, \underbrace{u_{n+1}, \ldots, u_{n+1}}_{p} \right\}$$

$$V := \left\{ \underbrace{v_1, \ldots, v_1}_{q}, \underbrace{v_2, \ldots, v_2}_{1_v}, \ldots, \underbrace{v_m, \ldots, v_m}_{1_v*}, \underbrace{v_{m+1}, \ldots, v_{m+1}}_{q} \right\}$$

where $1_u = p - c_u - 1$; $1_v = q - c_v - 1$ and $ncu = ((p-1) - c_u \cdot n + c_u + 1)$; $ncv = ((q-1) - c_v \cdot m + c_v + 1)$.

The control points d_{ij} can be used as parameter for the variation approach and lead to a linear system of equations.

If boundary curves are defined, they have to be approximated with the same degree, continuity and the same parametrization as for the surface in that parameter direction with the same variation principle. The resulting control points of this approximation can then be used to reduce the linear system of equations for the surface calculation.

To include an adjacent surface in the variational design process, the normal vectors of this adjacent surface have to be calculated along the boundary curve. If the surfaces have no common boundary, the boundary has to be projected onto the adjacent surface. Additional to the tangent plane information defined by the normal vectors, the tangent directions and the tangent lengths are needed. For this the variational surface can be calculated neglecting tangent conditions. From this first surface the tangents along the boundary can be calculated and then projected onto the tangent planes derived from the adjacent surface. With the discrete tangent informations a tangent curve can be calculated, that has the same degree, continuity and the same parametrization as the variational surface in that parameter direction. The resulting control points of this curve can then also be used to reduce the linear system of equations for the second variational surface calculation. Each curve, i.e. two boundary curves and two tangent curves, for each parameter direction of the surface has to be computed with the same knot vector. This can be done by a global automated subdivision method, which divides the curves of one parameter direction into several segments at the same parameter value.

For a successful variational design technique, an appropriate parametrization of the given points is necessary. As a first solution a mean plane can be used. After projecting the points onto this plane a parametrization can be calculated. Better results can be reached with the inclusion of the parametrization as an additional parameter in the variational design process.

A reparametrization

$$
\left.
\begin{array}{ccc}
u_k & = & u_{k_0} + \Delta u_k \\
v_k & = & v_{k_0} + \Delta v_k
\end{array}
\right\} \quad k = 1, \ldots, n_p
$$

included in the variational design process leads to the additional necessary conditions

$$
\frac{\partial}{\partial \Delta u_\ell} \sum_{k=1}^{n_p} w_{p_k} [F(u_k, v_k) - p_k]^2
$$

$$
= 2 w_{p1} < (F(u_\ell, v_\ell) - p_\ell), \frac{\partial}{\partial \Delta u_\ell} F(u_\ell, v_\ell) >= 0
$$

with $\ell = 1, \ldots, n_p$ and $< \cdot, \cdot >$ as the scalar product of two vectors.

The nonlinear system of equations combined with the linear system of equations from the variational design process can be solved with a Newton method. The appropriate initial values can be calculated with a mean plane approximation.

Examples

Figure 5: 80,000 digitized points (Dörrer & Broßmann)

Figure 6: Point set reduced to 8,000 points

Figure 7: Triangulated point set

Figure 8: Curvature scaled normals in data poins

Figure 9: Segmentation of data points

Conclusions

The structuring of a set of points, as a preparation for a variational surface design, uses a two dimensional triangulation. This algorithm is very stable but all points have to be projected into one plane. This disadvantage can be removed using a three dimensional algorithm.

Experiments with different degrees for the approximating function, used by the curvature estimation, turns out that lower degrees deliver the best results. One reason is that for higher degrees more points are necessary for the approximation and therefore the approximation is not local enough.

Acknowledgements

The presented variational design methods are implemented in the CAD/CAM system CATIA©in a joined co-operation development of HELLA KG Hueck & Co. in Lippstadt/Germany, TransCAT GmbH in Karlsruhe/Germany and the institute of Prof. Dr. Hans Hagen at the University of Kaiserslautern/Germany.

We would like to thank the companies Dörrer & Broßmann and Keiper GmbH for the courtesy of providing the data sets.

References

[Boi84]   Boissonnat, J.-D.: Geometric Structures for Three-Dimensional Shapes Representation, ACM Trans. on Graphics, Vol. 3, No. 4, 1984, pp. 266 – 286

[DeR92]   DeRose, T., Hoppe, H., Duchamp, T., McDonald, J. A., Stuetzle, W.: Fitting of Surfases to Scattered Data, SPIE Vol, 1830, 1992, pp. 212 – 220

[Ede87]   Edelsbrunner, H.: The computational geometry column, Bull EATCS, Vol. 31, pp. 111 – 114, 1987

[Ede94]   Edelsbrunner, H., Mücke, E. P.: Three-dimensional Alpha Shapes, ACM Trans. on Graphics, Vol. 13, No. 1, 1994, pp. 43 – 72

[Hag92]   Hagen H., Santarelli P.: Variational Design of Smooth B-Spline Surfaces, Hagen H. (ed.): Topics in Surface Modelling, 1992, 85-94

[Hag95]   Hagen, H.; Hahmann St.; Schreiber Th.: Visualization and computation of curvature behaviour of freeform curves and surfaces, Computer Aided Design, Vol. 27, No. 7, 1995, pp. 545 – 552

[Ham91]   Hamann B.: Visualization and Modelling Contours of Trivariate Functions, Dissertation, Arizona State University, 1991

[Hei97]   Heinz S.: Surface reconstruction and variable offset, Roller D., Brunet P. (ed.): CAD Systems Development - Tools and Methods, Springer1997, pp. 199 – 206

[San94]   Santarelli P.: Glättungskriterien und Algrorithmen zum Modellieren von Kurven und Oberflächen, Dissertation, Universität Kaiserslautern, 1994

[Sch97]   Schreiber, Th.: Approximation of 3D Objects, Computing Supplementum - Geometric Modelling, G. Farin, G. Brunnett (eds.) Springer, Wien 1997

[Sch94]   Schreiber Th.: Analyse und Approximation von unstrukturierten Daten und Freiformflächen, Dissertation, Universität Kaiserslautern, 1994

[Sch91]   Schreiber Th.: Voronoi Diagram based adaptive k-means-typed clustering algorithm for multidimensional weighted data, Lecture Notes in Computer Science 553, Bieri H., Noltemeier H. (eds.): Computational Geometry-Methods, Algorithms and Applications, 91, pp. 265 – 275

Computer Aided Spectacle Lens Design

Joachim Loos[1] and Günther Greiner[1]

The ability of the human eye to accommodate (i.e. to see both near and far objects sharply) decreases with age. In order to correct this defect so called progressive-addition lenses (PAL) have been developed. The desired properties of the PAL are expressed by curvature properties of the front surface of the lens. Thus optimal lenses are characterized by curvature constraints to the geometry. Construction of the geometry is performed using a penalty method. A suitable error functional, penalizing the deviation from the optical specifications has to be established. It is shown how to efficiently minimize this error functional iteratively. The combination of geometry reconstruction and optimization provides a useful tool for the main design loop of PALs. The system has been developed and implemented in close cooperation with the optical industry.

Introduction

In order to see objects at different distances sharply, the lens of the human eye is able to alter its curvature and thus its power of refraction. Our ability to accommodate, i.e. to change the shape of this lens, decreases with age, due to gradual stiffening of the lens. To properly correct this defect, known as *presbyopia*, we need glasses with low refractive power for viewing objects at far distances, and higher power for near distances. Thus, the required accommodation effort of the human eye is (partially) supplied by an increasing refractive power of the spectacle lens. Such a glass is called a *multi-focal lens* or a *progressive-addition lens (PAL)*. Usually, a PAL has a large *far vision area* with low refractive power in the upper part, and a smaller *near vision area* with higher power in the lower part. Moreover, the distribution of refractive power on the PAL should be smooth and monotonic.

The first PAL was manufactured in 1967 by B.Maitenaz [Mai67]. His approach is based on simple heuristics (versus exact computation) and makes

---

[1]IMMD IX, Graphische Datenverarbeitung, Universität Erlangen–Nürnberg
Am Weichselgarten 9, 91058 Erlangen, Germany

use of only a few free parameters. Another approach to PAL design computes a main curve with increasing mean curvature on the front surface. The whole surface is then defined by a conic which is moved along the main curve, also having increasing mean curvature from top to bottom. Free form surfaces have been used in a patent specification by Zeiss [Zei95]. However, rather than considering the surface as a whole, this technique only takes into account the location of some sample points when computing the PAL. When minimizing the errors at the sample points, common numerical optimization methods are used, which require large computational effort. In [Taz93] a variational approach is outlined. In contrast to our approach the functional to be minimized does not take into account the exact optical quantities It rather consists of six different terms, each being derived using simple heuristics.

We describe a new method for determining the geometry of PALs, which enables the PAL-designer to create a variety of different designs, and minimizes optical errors as much as possible. The computation takes into account the exact optical quantities of a lens and specific requirements of the designer (e.g. size and location of the near and far vision area).

The method described in the following has been developed and implemented in close cooperation with the optical industry. The shown results demonstrate the benefits of our approaches in practical applications. Moreover, lenses manufactured according to our method have proven to show a better error distribution than currently available progressive lenses.

Our presentation is organized as follows. We begin with an introduction of the relevant optical factors and derive an appropriate error functional for the surface of a PAL. Then we present algorithms for minimizing this functional.

The optics of progressive addition lenses

Characteristics of ophthalmic lenses

In ophthalmic lens optics, the characteristic quantity of a lens is its *vertex power*. Consider a lens with rotation axis l as in Fig. 1. l intersects the lens' back face at the *back vertex* B. An incoming parallel bundle of rays along l intersects l at the *focus* F. The lens' *focal length* is $|BF|$, the distance from B to F. Finally, the reciprocal value $P = \frac{1}{|BF|}$ is called the *(vertex) power*. It is specified in *diopter=dpt* $[\text{m}^{-1}]$. Note that for a diverging lens, the focus F will be virtual, being positioned in front of the lens (see Fig. 2). Therefore $|BF|$ is understood to be directed, being negative in the case of diverging lenses (see also [BW80]).

The refractive power of a lens is given by the *thick lens formula*

$$P = (1 - n)\kappa^b + \frac{(n-1)\kappa^f}{1 - d(1 - \frac{1}{n})\kappa^f} \tag{1}$$

where κ^f and κ^b are the curvatures of the front and back surface respectively. and d denotes the thickness of the lens, n being the index of refraction. Assuming a negligible thickness, we get

$$P = (n - 1)(\kappa^f - \kappa^b) \tag{2}$$

the *thin lens formula*. In the case that one of the surfaces, e.g. the front surface is non-symmetrical, there are two different principal curvatures and Eq. 1 and Eq. 2 have to be applied separately for each principal curvature direction. In that case, we get two refractive powers P_1 and P_2. The quantity $A := |P_1 - P_2|$ is called *astigmatism*. A non-vanishing astigmatism results in two focal lines at different distances. Obviously, a (symmetrical) eye, using a lens with $A \neq 0$, cannot accommodate on both focal lines. We require $A = 0$ for spectacle lenses[1].

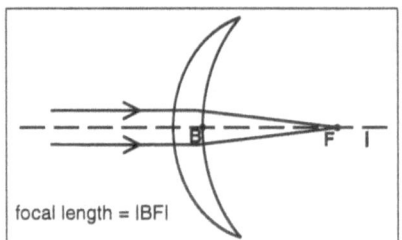

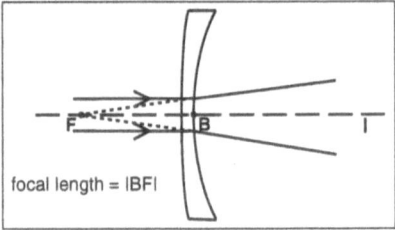

Figure 1: Focal length of rotationally symmetric converging lens

Figure 2: Focal length of rotationally symmetric diverging lens

Visual defects

An eye is said to be correctly sighted (emmetropic) if a point at infinity gets mapped sharply on the retina without any accommodation effort being made, i.e. when the lens is unaccommodated. In optical terms, a thin bundle of parallel light rays is deflected so as to converge to the same point on the retina (as in Fig. 3 a). We call an eye *short sighted (far sighted)*, if an incoming parallel ray bundle has its point of intersection in front of (behind) the retina (see Fig. 3 b, c).

---

[1] Currently, a possible astigmatic defect of a human eye will be corrected by an appropriate toric back surface of the lens.

These defects are corrected by spectacle lenses with negative or positive power of refraction resp. (see Fig. 3 d).

When looking at an object at a finite distance, the eye has to accommodate, i.e. the lens has to bend in order to increase its curvature and thus its power of refraction. With increasing age the eye looses part of its accommodative ability; it becomes *presbyopic*. In this situation uni-focal spectacles can only correct within a certain viewing range. Using a PAL with low refractive power in the upper part (far vision area) and higher refractive

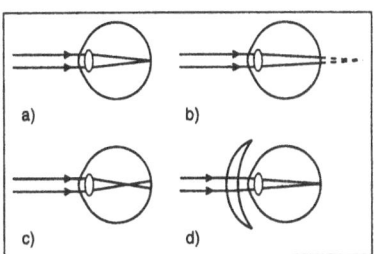

Figure 3: a) eye without visual defect b) far sighted eye   c) short sighted eye d) far sighted eye with correcting glass

power in the lower part (near vision area), the wearer can compensate the missing accommodation ability by moving the eye downward on the lens. An additional short or far sightedness is compensated as well, using a back face of appropriate curvature according to Eq. 2.

Requirements on progressive lenses

Due to the thin lens formula 2 we are able to separate front and back face of a lens. Assuming a spherical back face, we can therefore translate the requirements on a PAL into requirements on the curvature properties of an ideal front surface.

1. A certain mean curvature $P_f = \frac{\kappa_1 + \kappa_2}{2}$ is specified in the far vision area. In optical terms, the mean curvature is called *(surface) power*. Actually P_f could be chosen arbitrarily. However, flat or highly curved glasses are usually not desired, because of aesthetic reasons and due to optical errors caused by oblique astigmatism. Therefore P_f is often chosen in the range of 3 to 5 dpt[2].

2. The required surface power P_n in the near vision area is not specified directly. Instead, the *progression* $P_n - P_f > 0$ corresponding to the amount of missing accommodation will be specified.

3. Ideally, the front surface should be "locally spherical", i.e. $C := |\kappa_1 - \kappa_2|$ should vanish. In ophthalmology, $|\kappa_1 - \kappa_2|$ is called *(surface) astigmatism* or *cylinder*.

---

[2]Adjustment to the necessary refractive power, is achieved by an appropriate back surface of the lens.

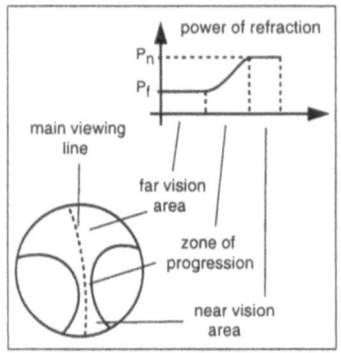

Figure 4: Distribution of refractive power on a PAL

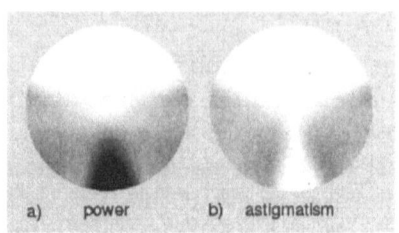

Figure 5: a) Refractive power. b) Optical error (astigmatism). White indicates lowest values for refractive power (a) and optical errors (b) respectively, black indicates maximal values.

Unfortunately, a surface being "locally spherical" (i.e. having vanishing astigmatism) has to be part of a sphere, i.e. it cannot have varying mean curvature (see also [Min63] for a quantitative statement). Consequently, we have to compute a surface which fulfills (1) and (2) and shows minimal astigmatism $|\kappa_1 - \kappa_2|$. Usually, one tries to get minimal astigmatism at the far and near vision area and on a *zone of progression* which connects both areas. The non-avoidable astigmatism should be reduced as far as possible and be placed near the boundary of the lens.

Figure 5 shows the front face of a typical multi-focal lens. The refractive power is displayed graylevel encoded on the left. White indicates lower, black higher values of power. The power increases smoothly from the upper far vision area to the lower near vision area. Optical errors (astigmatism) are shown on the right of 5. Errors at the far and near vision area and along the main viewing line are very small. Figure 4 shows the variation of mean curvature along the main viewing line. In the lower near vision part, the main viewing line is usually displaced into the nasal region of the lens, in order to take into account, that the eyes have to converge when looking at shorter distances.

Characterization of a PAL by an error functional

Representing the front surface of a lens by the graph of a function $f : \Omega \to \mathbb{R}$, with domain $\Omega \subset \mathbb{R}^2$ chosen as a disc, we can formulate the task of computing an appropriate front surface of the lens as a variational problem for the function f :

- The desired distribution of mean curvature on the graph of f is specified by a function $P : \Omega \to \mathbb{R}$.

- We specify weight functions $\alpha, \beta : \Omega \to \mathbb{R}$, $\alpha, \beta > 0$ for penalizing surface astigmatism and deviation of the prescribed mean curvature resp.

- The quality of the surface is measured by the following error functional J:

$$J(f) := \int_\Omega \alpha(\kappa_1 - \kappa_2)^2 + \beta(\frac{\kappa_1 + \kappa_2}{2} - P)^2 dS \qquad (3)$$

(We measure the error in the L^2-sense, since this allows an easier numerical treatment than e.g. the maximum norm.)

- The lens corresponding to the specified design parameters P, α and β is the solution to the following optimization problem:

> Determine f such that $J(f)$ is minimal.

The design parameter α, β and P are functions defined on the disc. The mean curvature function P is usually chosen to be constant along horizontal lines, and increases continuously from P_f to P_n. The function α specifies how much an occurrence of astigmatism is penalized by the functional. Similarly, β weights the deviation from the required mean curvature. Thus the designer specifies higher values of α and β in near and far vision area as well as along the main viewing line, whereas at the other parts lower values are used. Thus, the weight functions α and β provide good control over the optical quantities of the resulting lens (see Fig. 7).

Minimizing the error functional

In the last section, we introduced the functional

$$J(f) = \int_\Omega \alpha(\kappa_1 - \kappa_2)^2 + \beta(\frac{\kappa_1 + \kappa_2}{2} - P)^2 dS.$$

This functional is well-defined for functional surfaces having continuous partial derivatives up to order two. There is no hope to find an analytic expression of the solution in terms of α, β and P for the minimum of this functional. Only a numeric solution can be evaluated. We use a Ritz–Galerkin approach for the numerical analysis. Thus we consider a finite dimensional subspace V of $C^2(\Omega)$, e.g. V could be the space of bicubic tensor product B-splines. Then, we have to determine $f^* \in V$ such that $J(f^*) \leq J(f)$ holds for every $f \in V$.

Minimizing highly non-linear functionals like J using conventional numerical methods, e.g. the conjugate gradient method, is very time-consuming [MS92, WW92]. On the other hand, it is relatively easy to minimize a quadratic functional. In fact, minimizing a quadratic functional can be reduced to solving a linear system of equations. Therefore our strategy is to find local approximations to J that are quadratic. In addition, by iteration we will improve the approximations of J using quadratic functionals step by step. A similar method is described in [Gre94] where quadratic functionals are used to minimize certain curvature functionals.

The idea is as follows: Let us assume that we can find a family of quadratic functionals J_f, one for each *reference surface* f, such that $J_f(f) = J(f)$ holds. If the mapping $(f, g) \mapsto J_f(g)$ is continuous, it follows that J_f is a good local approximation to J in a neighborhood of f. We can therefore perform the following

Algorithm 1

1. Determine a rough approximation f^0 to f^*.

2. Do until convergence: Determine f^{n+1} by minimizing J_{f^n}.

If the approximating functionals J_f are chosen carefully, the sequence converges: $f^n \overset{n\to\infty}{\longrightarrow} f^*$. Now we present a quadratic approximation that turns out to be appropriate for our lens-design functional J.

Approximation with Taylor series. Since J is only dependent on the principal curvatures of the graph of f, only the first and second order derivatives of f have an influence on J. Thus, J has the following representation. $J(f) = \int_\Omega K(f_u, f_v, f_{uu}, f_{uv}, f_{vv})dudv$ with $K : \mathbb{R}^5 \to \mathbb{R}$. It is left to the reader, to derive an explicit representation for K. For simplicity we use $L(f)$ as an abbreviation for $(f_u, f_v, f_{uu}, f_{uv}, f_{vv})$. Then we can write $J(f) = \int K(L(f))$.

In order to achieve a quadratic approximation of J in the neighborhood of a certain reference function f^0, we use the Taylor expansion of K at f^0, omitting the terms of degree higher than two. So the quadratic approximation of J is given by

$$J_{f^0}(f) = J(f^0) + \int \mathsf{grad}\, K(L(f^0))^t L(h) + \frac{1}{2} \int L(h)^t\, \mathsf{Hess}\, K(L(f^0))L(h) \tag{4}$$

where $h := f - f^0$ and $\mathsf{grad}\, K$ and $\mathsf{Hess}\, K$ denote the gradient and the Hessian of K resp.. Obviously $J_{f^0}(f) = J_{f^0}(f^0 + h)$ is a quadratic functional in h and can therefore be minimized efficiently.

The following theorem shows that this approximation is well suited for minimization purposes.

Theorem 2 *Let f^* be an isolated non-degenerated minimum of J, i.e. $\frac{d}{dt}J(f^* + tg)\big|_{t=0} = 0$ and $\frac{d^2}{dt^2}J(f^* + tg)\big|_{t=0} > 0$ holds for every $g \in V$. Then Algorithm 1 produces a sequence $(f^n)_{n \in \mathbb{N}}$ that converges quadratically to f^* for every $f^0 \in V$ sufficiently close to f^*.*

Results

We found that our method leads to a fast converging sequence of approximative solutions.

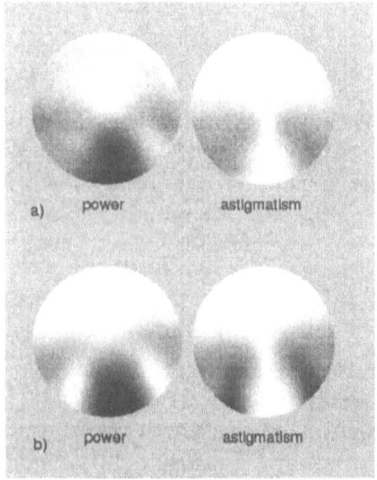

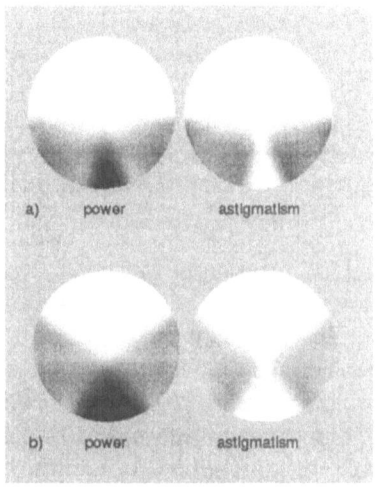

Figure 6: Examples of currently available PALs. Numerous defects are visible.

Figure 7: Multi-focal lenses, computed with the method introduced here

Obviously, our minimizing methods requires a sufficiently good approximation f^0 to the desired surface f^*, because only then we can expect convergence. In all cases we investigated, it proved to be sufficient, to simply use a flat disc $f^0 \equiv 0$ as the initial function. In order to reach the minimum faster, a sphere of radius $\frac{2}{P_f + P_n}$ can be used. Figure 7 and 5 show different PAL-designs that have been computed using the method described above. All PALs have a progression of 2 dpt. The computation took 5 seconds on a SGI MIPS R10000 195 MHZ processor. The solution is given as a 30×30 tensor product B-spline patches, trimmed to a disc. In order to compare the results of the optimization method to currently available lenses we displayed power and astigmatism of two progressive addition lenses in Fig. 6. The optical characteristics of these lenses are 3 dpt for the far vision area and 5 dpt for the near vision area.

The lens displayed in 6 a) has tolerable astigmatism, but the refractive power does not increase monotonicly from the upper to the lower part; the usable portion of the far vision area is therefore restricted to a small region. In contrast, the lens in 6 b) has correct refractive powers along the whole main viewing line. However, intolerable deviations occur next to the near vision area. Moreover, the techniques used to produce this lens could not sufficiently reduce the astigmatism. The lenses of Figure 7 and 5 obviously do not have these problems: the variation of the refractive power from far to near vision area is monotonic and the astigmatism is reduced to a satisfactory level.

Conclusion

We have presented a new method for the design and quality control of progressive-addition lenses (PAL). Using the thin lens approximation it is possible to reduce the design of a PAL to the optimization of its front surface, the back surface being fixed. This fact is important not only during the design process, but perhaps even more from a manufacturing point of view, since front and back surface are typically manufactured by different companies.

Optimization of the PAL's front surface is done by minimizing a suitably chosen error functional J involving only the principal curvatures of the front surface. We have presented a new efficient method to minimize this highly non-linear functional J by iteratively minimizing a sequence of quadratic approximations to J that are based on a reference surface. The construction of these quadratic approximations is based on Taylor expansion of J, and we proved quadratic convergence for the iteration. Examples show that the sequence of surfaces becomes quasi-stationary after two iterations. Efficient functional minimization is essential in many CAGD applications. We plan to further investigate how our method can be applied in a more general context in the future.

References

[BW80] M. Born and E. Wolf. *Principles of Optics*. MacMillan, New York, 1980.

[Gre94] G. Greiner. Variational design and fairing. In *Proc. EUROGRAPH-ICS '94*, volume 13. Eurographics, Blackwell Publishers, 1994.

[Mai67] B. Maitenaz. Image rétinienne donnée par un verre correcteur de puissance progressive. *Revu. Opt. Theor. Instrum.*, 46:233–241, 1967.

[Min63] G. Minkwitz. Über den Flächenastigmatismus bei gewissen symmetrischen Asphären. *Optica Acta*, 10:223, 1963.

[MS92] H. P. Moreton and C. H. Séquin. Functional optimization for fair surface design. In *Proc. SIGGRAPH*, volume 26, pages 167–176, 1992.

[Taz93] M. Tazeroualti. Designing a progressive lens *Curves and surfaces in geometric design*, pages 467–474, 1993.

[WW92] W. Welch and A. Witkin. Variational surface modeling. In *Proc. SIGGRAPH'92*, volume 26, pages 157–166. Siggraph, ACM SIGGRAPH, 1992.

[Zei95] Fa. Carl Zeiss. Gleitsichtfläche für eine Gleitsichtlinse. *European Patent Specification, Nu. 0 452 802 B1*, 1995.

Index